Solid and Hazardous Waste Management in Third World Countires

Solid and Hazardous Waste Management in Third World Countires

Daniel Dela Torre

SOCIETY Publishing

www.societypublishing.com

Solid and Hazardous Waste Management in Third World Countires

Daniel Dela Torre

Society Publishing

2010 Winston Park Drive,

2nd Floor

Oakville, ON L6H 5R7

Canada

www.societypublishing.com

Tel: 001-289-291-7705
001-905-616-2116
Fax: 001-289-291-7601
Email: orders@arclereducation.com

ISBN: 978-1-77361-282-9 (Hardcover)

ABOUT THE AUTHOR

Daniel Dela Torre

Daniel completed his MSc in Environmental Science from the University of the Philippines Diliman in 2016. His primary interests are ecological monitoring using remote sensing and the effects of climate variability on vegetation. He is engaged in the University of the Philippines as a Senior Science Research Specialist working on LiDAR mapping of agricultural resources.

TABLE OF CONTENTS

LIST OF GLOSSARY

Aerobic composting

Describes a method/ technique of composting the organic wastes by deploying bacteria that requires oxygen. The need is to have the waste in order to be exposed to air, either through rotating or by pushing the air through the pipes which would eventually then pass through the material.

Anaerobic digestion

Describes a method/ technique pertaining to composting which require no oxygen. This composting method leads to the production of methane and the method is also called as anaerobic composting.

Ash

The noncombustible by-products of solid in burning process.

Autoclaving

Sterilization done through a high temperature steam or a pressurized process.

Baghouse

An emission control device contained inside a combustion plant which also have an array of filters (fabric type) by which the flue gases penetrates in an incinerator flue. The particles are trapped and then prevents the gas from passing into the open.

Basel Convention

An agreement of international level on transboundary movement control of the hazardous/ harmful wastes and also for their proper disposal. This agreement was drawn up in March 1989 in the city of Basel, Switzerland. More than 100 countries endorsed the agreement as signatories.

Biodegradable material

Any sort of the organic material which could be processed in order to be broken down into simpler and more stable compounds by the microorganisms. Most of the organic wastes such as the paper and food etc. are known to be biodegradable.

Bottom ash

Relatively noncombustible, coarse, generally toxic residue of the burning that adds up on the grill of a furnace.

Bulky waste

Huge wastes, e.g. appliances, branches, trees, furniture, which could not be handled by using the normal MSW processing techniques.

Cell

The basic unit through which the development of landfill is done. It is also known as the general area where all the incoming waste is tipped, compacted, spread, and covered.

Chemocar

It is a specialized vehicle used for the sake of colleting toxic and the hazardous wastes from the shops, residences and other type of institutions.

Cleaner production

The processes developed for the sake of reduction of the wastes accumulated by production.

Co-disposal

The disposal of the multiple kinds of waste in one area of a dump or landfill. For example, the sewage sludge might be disposed of along with the standard solid wastes.

Cogeneration

The production of both the steam and electricity from one facility, using the same fuel source as well.

Garbage

This is for refuse in general, the things which are in everyday use. Some of the known MSWM manuals make use of the word garbage to refer to the "food wastes," although this usage is not that common.

Groundwater

It refers to the water present beneath the surface of the earth which fill ups the underground pockets that are also known as aquifers, and then that water is supplied to the wells and the springs.

Hazardous waste

It refers to the waste which is a kind of reactive, corrosive, and toxic or may be otherwise hazardous to the living things and/or also for the environment. For example, the majority of the industrial by-products are considered hazardous.

Heavy metals

The metals that are actually of some high density or high atomic weight, for example, the likes of mercury, cadmium and lead, which are known to be toxic for the living organisms.

Household hazardous waste

The products/ stuff that are used in residences, such as the cleaning compounds and paints etc. which are hazardous to all the living organisms and/or also for the environment.

Humus

Also known as compost, the end product of composting.

Incineration

The process of burning a range of solid waste which exist under the controlled kind of conditions in order to reduce its volume and weight and usually lead to produce energy.

Inorganic waste

The waste made up of the stuff which is other than the animal or plant matter, such as the likes of sand, glass, dust, and plenty of other synthetics.

Integrated solid waste management

It refers to the coordinated use of the management methods, organized within a set and there, each of which could play a role in running the overall MSVVM plan.

Landfill gases

The gases that originates from the decomposition of multiple types of the organic wastes, where the major types included are, the methane, hydrogen sulfide and carbon dioxide etc. Such type of mentioned gases may result in explosions at the landfills.

Landfilling

The final disposal of waste (solid) by using the technique of placing the waste in a kind of a controlled fashion over a place which is intended to be permanent. The Source Book refers to this term as in for both the sanitary landfills and the controlled dumps.

Leachate

It refers to the liquid (which could be partly generated using the decomposition of the organic matter) which has been seeped through a compost pile or a landfill and has been accumulated with the bacteria and several other types of the possibly toxic suspended or the dissolved materials. If this is uncontrolled then this could contaminate both the surface water and the groundwater.

Leachate pond

It refers to a tank or a pond that is constructed at a landfill in order to get the leachate from the vicinity. Normally, the pond used here is designed in order to provide some sort of known treatment of the leachate, by allowing for the settlement of the used solids or may be by through aeration for the promotion of the biological processes.

Lift

The completed layer referring to the compacted (compacted) contained in a cell at a landfill.

Liner

It is a kind of a protective layer, which is made up of soil and/or may be the synthetic materials and is installed along the sides and the bottom of a landfill in order to reduce or prevent the leachate's flow into the environment.

Manual landfill

It refers to a landfill where most of the operations are performed without the use of the mechanized equipment.

Market waste

It is primarily the organic waste, such as the leaves, unsold food and skins, which is discarded at or near the consumer food markets.

Mass-burn incinerator

It refers to a type of incinerator where the solid waste gets burned without any pre-processing or sorting.

Materials recovery

The obtaining of the stuff which could be be reused or recycled.

Materials recovery facility (MRF)

MRF refers to a facility for the sake of separating the commingled recyclables by the use of any manual or the mechanical means. Some MRFs are actually prepared for separating the recyclables from the mixed form of the MSW. MRFs eventually then bale out and goes for the marketing of the recovered materials.

Methane

It is an odorless, explosive, colorless and flammable gas, given by CH. It is produced through the anaerobically decomposing of the MSW at the landfills.

Microenterprise

It refers to a synonym for some small-scale enterprise/ business, which is normally family-based or may be a cooperative, which often employs less than 10 people and then may operate 'informally.'

Subsidy

It is the direct or the indirect payment given to a business, institutions or citizens by the government in order to encourage some desired activity.

Tipping fee

It refers to a fee paid for the dumping or unloading of the waste contained at a landfill, incinerator, transfer station or a recycling facility.

Tipping floor

It refers to an unloading vicinity for the vehicles, which are supposed to deliver the MSW to an incinerator or a transfer station.

Transfer

The act of transporting the waste material to a larger transporter vehicle from a collection vehicle.

Transfer point

It refers to a designated point, which is often located at the edge of a neighborhood, where the SMA collection transport vehicles are deployed to transfer the waste material to some larger vehicles for the sake of transport to the disposal sites.

Transfer station

It is a major facility where the MSW from the collection vehicles is consolidated in the form of loads, which are then transported by larger trucks or other means to more distant final disposal facilities, typically landfills.

Transition countries

The countries situated in the Eastern Europe and Soviet Union (former), which are working on the restructuring of their economies in various stages. The changes actually involve a shift away from being state-run considerably towards a number of novel configurations that ranges from moderate economic liberalization to a considerable dismantling of the role of state in the economy.

Vectors

The organisms which holds the disease-causing pathogens. At the landfills rodents, birds and flies are known to be the major vectors which trigger pathogens beyond the site of landfill.

Virgin materials

Any type of basic material/ stuff in order to be used with the industrial processes, which have not been used previously, for instance, wood-pulp trees, crude oil, iron ore and bauxite etc.

Waste characterization study

An analysis done of samples that are taken from a waste stream in order to find out its composition.

Waste collector

A personnel employed by some authority (local) or may be a private firm in order to collect wastes from the residences, community bins and the business sites.

Waste dealer

A middleman/ personnel who purchases the recyclable/ reusable stuff from the various waste generators and the itinerant buyers with the intention of selling the, but first apply some required sorting and processing, to the recycling industries or the wholesale brokers.

Waste management hierarchy

A ranking done of the operations of the waste management as per their energy or the environmental benefits. The need of the waste management hierarchy is to ensure that the waste management practices are found to be as environmentally sound as it is possible.

Waste picker

A personnel who finds out or collect the recyclables from the mixed waste wherever it could be disposed of or temporarily accessible.

Waste reduction

All means of reducing/ lessening the waste amount, which is produced initially and that must also be collected by the known solid waste authorities. This ranges from the product design and legislation to the local programs that are designed to have the recyclables and the compost-able away from the reach of the final waste stream.

Waste stream

The total waste flow that comes from a community, facility or a particular region.

Waste-to-energy (WTE) plant

A facility which employs the solid waste materials which could be either processed or raw, in order to produce energy. WTE plants have the inclusion of the incinerators which generates the required steam for the sake of district heating or for industrial use, or it may also produce electricity. They also have the facilities, which convert the landfill gas type to electricity.

Water table

The level situated below the surface of the earth and at which the ground surface becomes totally saturated with water.

Wetland

A vicinity which is flooded or is regularly wet and has a water table, which stands at a level similar or above the land surface for at least, considerable part of the year.

Windrow

It refers to an elongated pile of the aerobically composting stuff, which are periodically turned in order to expose the stuff to the oxygen and also for controlling the temperature for the sake of promoting biodegradation.

Working face

The length and width of the row where the waste is being accumulated at a landfill. This phenomenon is also called as the tipping face.

Worin castings

The stuff generated using the digestive tracts of the worms as they live on the surface of earth or the compost piles. The castings are known to be rich in nitrates, phosphorus, potassium, calcium and magnesium etc.

Worm culture

A relatively cooler type of the aerobic composting process, which uses the microorganisms and worms. This phenomenon is also known as venniculture.

Yard waste

Leaves, pruning, grass clippings and several other types of the natural organic matter which are removed from the gardens and yards.

Airspace

The projected type of Bank Cubic Yards (BCY) belonging to the landfill which is required to be filled with the waste material as that is determined by the survey conducted and/or by using some other types of the known engineering techniques.

Baler

A piece of gear that is used for compressing and to form some of the recycled material into bales.

Boiler (Wheelabrator)

It refers to a device that is used for absorbing the available heat which is released during the time of the combustion process of the burning waste. This combustion generates steam, which could be either sold or may be converted into a source of electrical power.

Brownfield Development

It is defined as an "abandoned, under-used or idled commercial and industrial facility where the redevelopment or expansion is considered to be complicated by perceived or real environmental contamination." The Waste Management has been known to have worked with communities, states and several other types of the economic development stakeholders in order to pursue a large number of brownfield development endeavors, all over the U.S.

Capping

It refers to the process of final cover material placement over the landfill.

Cell

Landfills are known to be built in phases, known as cells, which then adjoin with one another, detached by a berm for having leachate within a vicinity. The entire allowable vicinity would be alienated into distinct cells for the sake of construction.

Closed Site (Landfill)

A landfill, which is known to reach the level of its allowable waste material capacity and has also been eternally certified and coated as closed by the authorized state regulatory agency.

Closure

The definite period of time after which a landfill has reached its allowable capacity but that too before it has got the closure certification from the authorized state regulatory agency. At the time of the closure period, several known activities must be done in order to comply with the environmental and several other regulations such as capping and landscaping etc.

Commercial Customer

A part of the business, which is made from an industrial and commercial collection.

Construction and Demolition (C&D)

A waste material stream, which is mainly received from different types of construction sites. Some of the examples of such types are the C&D waste and that include, but are not limited to, concrete, paneling, carpet, rebar, wood and linoleum.

Container

Any container used for the sake of collecting the waste from residential, industrial and the commercial locations. The containers vary in type and size as per the needs of the customer or may be as per the restrictions of the community. Containers are also sometimes, known as the dumpsters.

Chemical Waste Management (CWM)

This term refers to the operating name given to the Waste Management's hazardous landfills. WM presently operates and owns 5 toxic waste landfills in the U.S.

Landfill, Municipal Solid Waste (MSW)

A landfill, which has been allowed by a state regulatory agency for collecting the municipal solid waste. This kind of landfilling should have the design and the properties that features some of the specifics to such type of landfill, which have been proven by the state regulatory agency.

Landfill footprint

Parcels of the land, which are permitted and designated in order to perform the landfilling activities. This would include the likes of entrance, buffer area, staging area and the area, which would be accepting the waste for the sake of disposal (the waste footprint area).

Leachate

Liquids which have come in close approximation with the waste. Leachate adds up in the waste material footprint of the landfill. Leachate levels contained in the landfill should be monitored and this should not exceed the given state regulatory agency levels. Depending upon the nature of the site, there are plenty of distinct ways to handle accumulated leachate. Some of these are given as below:

Collecting Leachate in the tanks and transporting it off-site periodically for the sake of disposal and waste treatment

Accumulating the Leachate in the evaporation ponds which eventually enables it to naturally evaporate into the atmosphere

Discharging the Leachate into the sewerage system

Recirculating Leachate back into the landfill in order to support in the waste's biodegradation.

Recycle America Alliance (RAA)

Recycle America Alliance (RAA) refers to a consolidated and the majority owned subsidiary of the Waste Management, Inc. RAA handles in excess of some eight million tons of the commodities every year. It operates nearly 80 recycling plants and also give the marketing responsibility locations in excess of 180 in the U.S. and Canada region. Moreover, it operates seven different types of the container processing facilities as well, where one is the plastics recycling facility, and four other are the electronics recycling facilities.

Residential Customers

A part of the waste collection business, which is made from a single and multi-family dwellings.

Route

An explicitly focused way, which a driver follows and has been designed for the sake of improving efficiency and also to provide the most optimal service to the customers.

Scale House

It could be found at either a transfer station or a landfill. It is the office which is situated at a short distance away from the main entrance and at this point, all of the incoming vehicles should stop in order to be measured or weighed and also to receive a disposal ticket.

Solid Waste

"Regular" garbage collected from the non-industrial sources, for example, from the residential homes, retail centers, restaurants and the office buildings. Some typical types of the MSW includes the likes of discarded food items, paper and several other types of the general discards. T green waste is actually taken as the MSW and includes things such as the yard clippings, trees and leaves etc.

Special Waste

Any type of waste material that requires some special sort of handling. Special waste is non-toxic waste, which generally comes from an industrial generator and it should also be profiled in order to ensure that it must not contain any form of the elevated levels of the potential toxic materials or the chemicals.

Tipping fee

It refers to the fee paid by anyone who disposes waste at a landfill.

Transfer Station

A facility, which has a large pad where the commercial and residential collection vehicles empty the waste material of their trucks. Other machinery, such as the bulldozers are then used in order to push the garbage into the long-haul trailers for the sake of transporting it to the disposal facilities.

Turbine Generator (Wheelabrator)

It refers to a device, which transforms the steam's heat energy from the boiler into the electrical power.

Waste Stream

Particular types of the waste material found in the customer's disposal such as the trash, metal, cardboard and aluminum etc. or may be a broader definition of the disposal type such as the MSW, Hazardous and C&D etc.

Working Face

The segment of the landfill where the waste material is being placed actively by the disposal vehicles.

A

Aerobic - Life or may be the processes which require, or are potentially not destroyed by with the existence of oxygen.

Aerobic Treatment – It refers to a process through which the microbes begins to decompose the complex type of the organic compounds where there is some presence of oxygen and use the unconventional source of energy for the sake of reproduction and growth. (Such type of processes also include enhanced aeration, rotating biological contractors and the trickling filtration.)

Agricultural Waste - Poultry and the cattle manure, and also the residual stuff present in the liquid or solid form, which is produced from the production and marketing of livestock, poultry or the fur-bearing animals, which also includes vegetable, grain and the fruit harvest residue.

Aerated static pile – Forced ventilation technique of composting where a free-standing type of a composting pile is ventilated using a blower passing air through the perforated pipes which are situated right beneath the pile.

Algae – Simple type of the rootless plants, which grow in the sunlit waters in the proportion to the available nutrients amount. Algae can adversely affect the quality of the water by lowering the amount of the dissolved oxygen contained in the water. Algae are actually the food as well for fish and several other types of the small aquatic animals.

Algal Blooms - Sudden sprays of algal development that can adversely affect the quality of water and indicate all of the potential toxic changes in the water (local) chemistry.

Aquifer – A geologic type of a formation, group of formations or may be a part of some formation, which contains enough saturated and the permeable stuff in order to yield considerable quantities of water to springs and wells.

Asbestos - A mineral fiber, which has the potential to pollute water or air and could result in cancer or asbestosis when this is inhaled. EPA has severely restricted or completely banned its use in the construction and manufacturing.

B

Bacteria – It is a microscopic living organisms, which can aid in the pollution control through the metabolizing process of the organic matter in the oil spills, sewage or other pollutants, However, the bacteria present in soil, air or water could also cause human, plant and animal some health issues.

Bale – A bound and a densified cube of the recycled stuff, for example, waste paper, rags or the scrap metal etc.

Bulky Waste – It refers to the large chunks of the waste stuff, for example, appliances, trees, furniture, large auto parts and stumps etc.

Buy-Back Center – This concept is also known as the Redemption Center. It is a facility where actually the groups or even individuals bring their recyclable wastes in return for payment.

C

Cap – It refers to a layer of clay, or may be some other impermeable sort of the material, which is installed over the top of a landfill (closed) in order to prevent any sort of entry of possible rainwater and also works on to minimize leachate.

Compaction – It is the reduction of the bulk (solid) waste material by the act of tamping and rolling.

Compost - A humus or may be a soil-like material which is created from the source of microbial decomposition, aerobic of the organic materials, for example the food scraps, manure and yard trimmings etc.

Compostable - Capable of undertaking biological decomposition in some compost site as part of some available program, in such a way that the material used is not distinguishable (visually) and also breaks down into the form of carbon dioxide, inorganic compounds, water and biomass at a rate that is consistent with the known compostable materials.

Compostable Plastic - A plastic material, which undergoes some sort of degradation experience by using the biological processes during its composting in order to yield CO2, inorganic compounds, water and biomass at a rate that is consistent with the other known compostable materials and eventually leave no visible, toxic or distinguishable residue.

Composting – It refers to a process of some accelerated form of the biological decomposition of the organic material which is used under controlled conditions.

Composting Facility:

An offsite facility which contains the organic component of the municipal waste material (solid type) is decomposed under some known controlled conditions

An aerobic process where the organic type of waste materials are grounded or shredded and then they are lead to be decomposed in order to humus in the windrow piles or may be in the drums, mechanical digesters or similar type of enclosures.

Conservation - Renewing and preserving, whenever possible, the natural and the human resources. The use, improvement and the protection/ maintenance of the available natural resources as per the principles, which will ensure their highest possible social or economic benefits.

Construction and Demolition Waste Processing Facility – A volume reduction plant or a facility, the operations of which solely include the decrease in the volume of structure and flattening of the waste material generated elsewhere.

Contaminant – The contaminant phenomenon refers to any physical, biological, chemical, or radiological substance or matter, which has an adverse effect on air, soil or water.

Contamination - Introduction into air, soil or water of the microorganisms, toxic substances, chemical/ chemical wastes, other wastes, or wastage of water in a concentration, which makes the medium flabby for its following intended use. This concept also applies to the surfaces of buildings, objects and various other types of the household and the agriculturally used products.

Contract Collection – A collection that is made by a private collector under a formal level of agreement along with the concerned municipal authority where the duties and the rights of the respective parties are set forth.

Corrugated cardboard – It is a cardboard that comes with corrugations (This is something which can also be glued in order to flat the cardboard on either one or both of the sides)

Cost/Benefit Analysis – It refers to an evaluation (quantitative) of the costs that would have incurred by the process of implementing some known environmental regulation vs. the overall benefits that are available to the society of the proposed action.

Cover Material – Refers to the soil which is used for covering the compacted solid waste material in a sanitary landfill.

Cradle-to-Grave or Manifest System - A procedure where the toxic waste materials are identified and then worked upon just in the way they were produced, transported, treated and disposed of through a series of linkable, permanent, descriptive documents, such as the manifests. The phenomenon is also very commonly referred to as the cradle-to-grave system.

Crumb Rubber - Ground rubber separates the size of silt or sand that is used in the plastic or rubber products, or they are further processed into the reclaimed form of the asphalt or rubber products.

Cullet – Some waste or broken glass that is suitable for the purpose of re-melting.

Curbside Collection – Technique used for the purpose of collecting the recyclable/ reusable waste materials at homes, businesses or the community districts.

D

Decomposition – The process pertaining to the breakdown of matter through which fungi and bacteria, changing the chemical properties and the physical appearance of the used materials.

Deconstruction - A technique that different practitioners are deploying in order to recover some valuable type of building materials, decrease the amount of waste material that they send to landfills and also work on under the concept to mitigate other known or unknown types of the environmental impacts. It is the actually the disassembly of some building in common terms and also the recovery of its materials, which is often sometimes thought of as a construction done in reverse.

Decontamination – Technique used for the removal of the harmful substances such as the toxic chemicals, harmful bacteria or several other types of the harmful organisms, or the radioactive material that are coming from the exposed rooms, individuals and furnishings in buildings or may be onto the exterior environment.

Disposables – It refers to the consumer products, several other items and the packaging used, at least once or may be a few times and discarded.

Disposal – The final placement or demolition of hazardous, radioactive, or other types of the surplus, waste materials, or banned pesticides or other chemicals, polluted soils and the drums containing all the toxic/ hazardous materials from the required removal actions or through the accidental releases. Disposal of the waste materials may be accomplished through use of some approved type of the surface impoundments, secure landfills, land farming, ocean dumping, deep-well injection, or incineration.

Diversion Rate – The percentage of the waste materials which is actually diverted from the customary disposal such as incineration or landfilling to be recycled, re-used or composted.

Drop-off Center – A type of a central, predesignated area, facility or a building, which is set up in order to receive the recyclables that are dropped off by the individuals.

Dual Stream – A kind of a collection system where the reusable or the recyclables are channeled into two streams, so that the usual one type is the paper/fiber and the other one is a container.

E

End User – The consumer of the products for the sake of recycling. This actually excludes the products used for re-use or combustion for the purpose of energy recovery.

Energy Recovery – The act of obtaining energy from the waste material using a wide variety of processes, such as combustion.

Extended Producer Responsibility – The environmental protection strategy that is employed to reach a desired environmental objective related to the impact of the decreased total environmental from a product, by nominating the manufacturer of the intended product responsible for the complete life-cycle of the product and mainly for the take-back, final disposal and recycling of the product.

F

Ferrous Metals – The magnetic metals, which are derived from steel or iron. The products that are made from ferrous metals include the likes of appliances, containers, furniture and the packaging materials like the steel drums and barrels. The recycled products also involve the processing of the tin/steel cans, strapping, and metals that are derived from the appliances into the novel products.

Fill - Human-made deposits of some natural soils types or the rock products and also the waste materials.

Fly Ash – It is the non-combustible type of the residual particles which are known to be expelled by the flue gas.

Food Chain – It is the known sequence of the organisms, where each of which bring into use one of the next and the lower member of the sequence as a definite food source.

Food Processing Waste – The food residues which are produced during the industrial and the agricultural operations.

G

Garbage – It refers to the vegetable and animal waste material which result from the handling, sale, storage, preparation, food cooking, and the serving of foods.

Gaylord – it refers to the trade name that is used for a large size of a reusable crenelated container, deployed for the shipping materials (dimensions are approximately 40 x 48 x 37 inches)

Glass Containers - For the purpose of recycling, the containers like jars and bottles for cosmetics, drinks and food and several other similar or dissimilar products. When these materials are being recycled then the container glass is usually separated into different types of other color categories for the purpose of converting them into new containers, fiberglass insulation or for the construction materials.

Grass-cycling – It refers to the source reduction kind of activities where the grass clippings are left behind on the lawn/ garden after mowing.

Green Remediation – The practice of taking into account all of the environmental effects related to the incorporating options and the remedy implementations in order to maximize the total environmental benefits related to the cleanup actions.

Greenhouse Effect - The phenomenon of warming of the atmosphere of the Earth is credited to a buildup of carbon dioxide or several other types of known gases. There are some scientists who are of the opinion that this kind of a build-up allows the rays of the sun to actually heat the Earth, while at the same time, making the infra-red type of the radiation atmosphere impervious to that of the infra-red radiation, thereby actually preventing any possible chance of counterbalancing the loss of heat.

H

Halogen – It refers to a type of some incandescent kind of a lamp that is known to have higher energy-efficiency as compared to the standard ones.

Hauler – It refers to the garbage collection company, which offers complete/ comprehensive refuse removal service, where many of such companies will also collect the recyclables.

High Density Polyethylene (HDPE) – A type of plastic which used for making the milk jugs and several other types of the rigid plastic bottles. It is also that material which is used for making the plastic bottles and multiple other products, which produces the toxic fumes when they are burned.

I

Incinerator – It is a kind of furnace for the purpose of burning waste under the controlled conditions.

Indirect Emissions – Emissions that are associated with the ingestion of acquired electricity, steam and heating or cooling using a facility or entity.

Industrial Pollution Prevention – Having a combination of some industrial source decline and hazardous chemical use as a substitution.

Industrial Process Waste - Residues from the products/ materials that are collected at the time of the manufacturing operations.

Industrial Waste - Unwanted type of the waste materials produced from an industrial operation, which may be liquid, solid, sludge or toxic waste.

Inorganic Chemicals – They are the chemical substances of some sort of a mineral origin, which are actually not of basic type of carbon structure.

Institutional Waste – It refers to the waste that is generated at the different type of institutions such as schools, hospitals, libraries and prisons, etc.

Integrated Pest Management (IPM) - A mixture of some sort of chemical and other type of non-pesticide methods used to control pests.

Integrated Waste Management - Deploying a variety of different practices in order to handle the municipal solid waste, which can include the source reduction, incineration, recycling and landfilling.

L

Leachate Collection System - A system that is used to gather leachate and then eventually pumps it to the surface for the purpose of treatment.

Leaching - The act by which the soluble type of the constituents are dissolved and then filtered through the soil by using an infiltrating fluid.

Life Cycle of a Product – It involves all the stages of the development of a product which goes from the extraction of fuel for generating power for production, marketing, use, and its disposal.

Liner

A comparatively resistant barrier which us designed in order to keep the leachate intact inside a landfill. The liner materials may include the dense clay and plastic.

A sleeve or an insert used for the sewer pipes in order to prevent the infiltration or leakage.

Litter

The highly perceptible lot of waste (solid) which is carelessly discarded other than the normal garbage and the regular trash collection and the disposal system.

Twigs and the leaves which are fallen from the trees of the forest.

M

Mandatory Recycling – The type of programs that by law require some consumers in order to separate the trash so that some or may be all of the available recyclable materials are eventually recovered for the sake of recycling rather than opting for the incinerators or the landfills.

Microbial Growth - The multiplication or the amplification of the microorganisms, for example, the likes of bacteria, diatoms, algae, plankton, and the fungi.

Minimization - A widespread program in order to minimize or may be completely eliminate the waste materials, which are normally applied to waste materials at their point of origin.

Mixed Glass – It refers to the recovered type of the container glass which is not actually sorted into the categories, for example, color or grade etc.

Mixed Metals - Recovered metals which are not sorted into any of the known categories such as the aluminum, steel cans, tin or the ferrous or the non-ferrous metals.

Mixed Paper - Recovered paper material which is not actually sorted into the categories such as the old magazines, old corrugated boxes or the old newspapers etc.

Mixed Plastic - Recovered plastic material which is unsorted by the category.

Municipal Collection – Solid waste material collection taken from all the residents, thereof through a municipal authority.

Municipality – Any town, borough or city situated within the state.

N

Newspaper – Used or may be the discarded newsprint that has a minimum amount of contamination contained by the food or other types of the known material.

Nitrogenous Wastes – Vegetable or the animal residues, which has a considerable amount of nitrogen.

O

Open Burning - Uncontrolled fires present in an open sort of a dump.

Open Dump – It refers to an uncovered site, which is used for the purpose of disposal of the waste material without having any environmental controls.

Organic

Referring to or may be derived from some of the known living organisms.

In chemistry, it refers to any of the compound that contains carbon.

Organic Chemicals/Compounds - Naturally existing (plant-produced or animal or synthetic) substances that comprise mainly of the carbon, nitrogen, hydrogen and oxygen.

Organic Matter - Carbonaceous type of the waste material that is contained in animal or plant matter and which is known to be originating from the industrial or the domestic sources.

Organism - Any form of the plant or the animal life.

Ort - A small scrap or perhaps the leaving of food, right after a meal is completed. This term is usually used in the plural.

Other Ferrous Metals – The recyclable/ re-usable metals collected from strapping, metal and furniture that is found in consumer electronics and tires, but does not actually include the metals, which are found in the construction materials or cars, ships and the locomotives.

P

Packer truck, compacter truck – A refuse or the recycling collection automobile, which actually works to compact the waste materials. This also is the most usual type of the refuse collection automobile used.

Pallet – A type of a wooden platform that is used in linking with a forklift vehicle for the purpose of moving the bales or any other type of the large items. Pallet is also sometimes called as a ‘skid’.

Paper - In the recycling business, the term paper refers to the materials and the products, including newspapers, office papers, magazines, corrugated containers, paper-bags and some paperboard type of packaging, which can be recycled into new form of the paper products.

Paper Processor/Plastics Processor - Intermediate type of a facility where actually the recovered plastic or paper materials and products are sorted, prepared and decontaminated for taking then for the final recycling phase.

Participation Rate – The portion of the population, which is contributing in a recycling program.

pH – It refers to an expression of the intensity of some sort of an acid or basic condition of a liquid material, which may range anywhere from 0 to 14, where 0 refers to the point of the most acid and 7 refers to a neutral value in this regard. The natural water is known usually to have a pH value between the range of 6.5 and 8.5.

Point Source - A stationary site or may be a fixed facility from which the pollutants are discharged. It also refers to any single point of the identifiable source of pollution, for example, a pipe, ship, ditch, ore pit or a factory smokestack etc.

Polyvinyl Chloride (PVC) - A tough and an environmentally imperishable plastic material, which releases the hydrochloric acid when it is burned.

Post-Consumer Recycling – It points to the use of some materials, which are generated from the consumer and the residential waste material for the sake of some novel or similar purposes, for example, conversion of the wastepaper collected from the offices into the corrugated boxes or the newsprint.

Pre-Consumer - A term that is used for describing the material, which is being reused/recycled before the time, it ever goes to the market, for example, paper scraps off of a paper factory floor that is going back into the next batch of paper. The waste material that is generated during the manufacturing process.

Pre-Consumer Materials/Waste - Materials that are generated in the manufacturing and the conversion processes, such as the manufacturing scrap and cuttings and trimmings. This also includes the print overruns, over issue publications, and also the obsolete inventories.

Pressed Wood Products - Materials that are used in the furniture and building construction, which are made from the wood veneers, fibers or the particles that are bonded together with an adhesive under pressure and heat.

Processed Wood – The recycled wood or the treated wood or may be any combination thereof that actually has been processed at some volume reduction facility.

R

Reclamation – In recycling, the process of the restoration of the waste materials collected from the waste stream which are taken into a beneficial use and that may be for the sake other than that of the original use.

Recovery Rate – The percentage of the usable recycled waste materials, which have been eliminated from the net amount of the municipal waste (solid) material that is generated in a particular area or may be by a particular business.

Region – Two or may be more municipalities that have been linked together by making a district or signing an agreement (inter-local) or perhaps signing some sort of a mutual contract for a definite time period, concerning waste (solid) management within such kind of the municipalities.

Regional Authority – The administrative body that is being delegated the responsibility of waste (solid) material management for two or may be more municipalities that have been linked together by creating a district or perhaps signing an agreement (inter-local) or may be signing a mutual contract for some definitive period of time.

Regulated Wood Fuel – It refers to the processed wood from that of the demolitions and the construction activities which has been sorted in order to eliminate the plastics, gypsum wallboard, plaster, asbestos, wood and the asphalt shingles that contains creosote or to which the pesticides have been pragmatic or which has the substances defined as toxic.

Renderer – Some known entity who renders and that could be a business or an individual.

Residential Waste – It refers to the waste material generated in some single and also the multi-family homes, including the likes of the newspapers, disposable tableware, clothing, food packaging, cans, tins, bottles, food scraps, and the garden trimmings, which are other than those that are averted to the courtyard composting.

Resources Recovery Facility – A type of a facility that utilizes the processes in order to reclaim the energy gained from the municipal waste (solid).

Reuse - Using a component or a product of the municipal waste (solid) in its known original form that is done more than once, for example, the refilling process of a glass bottle, which has been returned or perhaps using a coffee can in order to hold bolts and nuts.

Rubbish – This includes the waste (solid), which excludes the food waste material and the ashes produced from homes, workplaces and the institutions.

S

Salvage – The term refers to the exploitation of the waste materials.

Signal Words - The words, which are used on a pesticide label, such as Danger, Caution, Warning in order to point out the level of toxicity.

Single Stream Collection – It refers to a collection system where actually the recyclable types are completely commingled, containers and the mixing fiber (papers).

Sludge Processing Facility – It refers to a type of a volume reduction plant, where the operations performed, involve completely the decrease in volume of the treatment of water, treatment of sewage or the industrial sludge that is generated elsewhere.

SMART – the literal meaning of the term in the context id Save Money And Reduce Trash – This is another name given to the concept of Pay-As-You-Throw – These are actually the systems under which the residents pay for the municipal waste material management system and the disposal services by means of volume or weight that is collected. This does not have a fixed fee.

Solid Waste Planning Region – Those type of the municipalities that are operated well within the defined boundaries of the planning agencies (regional) or as prescribed in the state waste (solid) management plan.

Source Separation – It refers to the segregating various types of the waste materials at the generation point, for example, the separation of paper, glass and metal from the other waste materials in order to make the process of recycling more efficient and simpler.

Sustainability – In order to meet the needs of the current time and that too, without any compromising of the ability of any of the future generations for meeting their own needs.

T

Toxic Substance – A mixture or chemical, which may expose to an unreasonable risk of some sort injury to environment or to health.

Toxic Waste – It refers to some waste material, which can lead to some sort injury if swallowed, inhaled or absorbed through the skin of human or animal.

Trash – Waste material which is considered offensive or worthless and that is thrown away. This term is also generally defined as a dry sort of waste material, but in the normal usage it refers to a synonym for garbage, refuse or rubbish etc.

Treated Wood – It is the type of wood that contains an adhesive, stain, paint, fire retardant, preservative or pesticide etc.

U

Used Oil – It is the motor oil which is spent in the trucks or the passenger cars and is collected at some specified locations for the purpose of recycling (this does not include in the municipal solid waste category).

V

Vector

An organism that is often some rodent or insect and which is known to carry diseases.

Plasmids, bacteria or viruses, which are used in order to transport the genes into a cell (host). A gene is actually something which is placed in the vector and the vector then 'infects' the bacterium.

Vermicomposting – It refers to the process through which the worms convert the waste material (organic) into the form of worm castings such as the fertile, dark, granular excrement of a worm. Castings are known to be very rich in the plant nutrients.

Volume Reduction – It refers to the processing of the waste materials in order to decrease the amount of space that they occupy and this is usually done by compacting, incineration, shredding, or composting.

W

Waste:

Unwanted or undesired stuff left over at the end of some manufacturing process.

Refused material found from the places of animal or human habitation.

Waste Characterization - Identification of chemical and microbiological constituents of a waste material.

Waste Exchange – A kind of arrangement where the companies exchange their collected waste material for enjoying the benefits of both parties.

Waste Feed - The intermittent or may be the continuous wastes' flow into an incinerator.

Waste Generation - The volume or weight of the waste materials and product, which penetrate into the waste material stream before the initiation of the processes such as the recycling, combustion, composting or landfilling takes place. Waste generation can also represent the total amount of the waste material which is generated by any given source or may be a category of sources.

Waste Minimization - Measures or techniques that reduce the amount of wastes generated during industrial production processes; term is also applied to recycling and other efforts to reduce the amount of waste going into the waste stream. See also Waste Reduction.

Waste Stream – It refers to the total flow of the waste (solid type) from homes, institutions, businesses and numerous manufacturing plants, which is recycled, disposed, burned in the landfills or the parts thereof such as the "recyclable waste stream" or the "residential waste stream"

Waste-to-Energy Facility/Municipal-Waste Combustor – A facility where the recovered type of the municipal waste material (solid) is transformed into some usable form of energy, which is usually through the process of combustion.

Waste Reduction – Effort put in the reduction of the amount of waste materials that is made to enter the waste stream through a particular source by using the redesigning products or the known patterns of consumption or the production, such as, using the returnable/ disposable beverage containers. The term is synonymous with Source Reduction.

Windrow – The term refers to a long, relatively narrow, and low compost pile. These have a very large exposed area (surface) that supports for the drying and the passive aeration

Wood Packaging - Wood packaging refers to the wood products, for example, pallets, barrels and crates etc.

Y

Yard Trimmings – Leaves, weeds, grass clippings, branches up to one inch in diameter and then the pruning from the gardens and the yards.

Z

Zero Waste – The term refer to the Zero Waste capability which is actually a goal that is both visionary and pragmatic in order to guide the people for emulating some sustainable natural cycles and where all the discarded materials are actually the resources for several others to bring into any use. Zero Waste actually means the designing and managing of the processes and the products in order to reduce the toxicity and volume of waste and the other such materials, which are then to be conserved and eventually recovered from all the resources, and not to burn or bury them. Implementing the Zero Waste mechanism will remove all the discharges to land, air or water, which may be a threat to planetary, animals, humans or the plant health. (Zero Waste International Alliance)

LIST OF FIGURES

PREFACE

Any unwanted material that turns out to be toxic, radioactive, explosive, irritating to the respiratory system, carcinogenic and has a negative effect on the environment as well as harmful to human life is called hazardous waste. Developing countries are quickly becoming industrialized and the number of factories is increasing. One common thing in all of these countries is the creation of waste from in huge quantities from industries, hospitals, as well as households. This waste often becomes hazardous easily due to poor solid waste mechanisms put in place. This in turn poses a great danger to the life of humans, animals, and the environment with several cases of injury and sometimes death being reported.

Mostly in these developing countries, the main types of hazardous waste are industrial and medical waste. However, there is also a significant amount of household waste being produced daily. Industrial waste is mainly from the numerous processing and manufacturing factories which release solid, liquid and gaseous waste daily. Medical waste is mainly from hospitals and health facilities which is increasing with more people seeking medical care every day. Back at their homes the people also create large amounts of waste. And despite there being designated dumping areas, crude dumping is a norm which increases risks. Some factories also send their liquid waste directly to rivers without treating it. Others release gases and smoke into the atmosphere which increases the effects of global warming.

Interestingly, there are laws and policies that govern waste management in these developing countries. However, due to policies not being implemented well, lack of resources and sometimes ignorance by the people, the risks arising from poor waste management are always prevalent. Any waste should be well controlled from the point of production and well stored until it's transported to the ultimate point of disposal. This is not always the case in developing countries whose waste disposal methods being unregulated and unconventional. For example, the owner of a household may not know about segregation of non-hazardous and hazardous wastes. He or she therefore puts all waste in one container and maybe burns it, not knowing there was an explosive waste in the batch. This of course exposes people to more risk. Health facilities have incinerators and pits where they securely dispose all their waste. However, .investigations show that not all of them carry our proper disposal of their waste posing much danger to the environment and life.

Despite legislation being in place to control waste management, it seems there needs more to be done to not only reduce the amount of waste being produced but also to manage it and make it non-hazardous. This includes education to the population, segregation of waste, taking tough measures on all those who disregard these policies, allocating more resources to the waste management departments and inclusion of technology as a means of waste management.

The technology in developing countries may not be as effective as it is in the developed countries, but with proper use of the current one available to them can make a huge difference. There are sites marked as waste disposal sites, however, all these lie misused where you find them filled to capacity and spilling over but not being emptied. With flies buzzing all over and a strong stink polluting the air, this is very hazardous. However, with the use of innovative technology, this is easily manageable. Through the use of conventional ways of disposal such as incineration, landfilling or even treating waste in factories before releasing it, great strides can be made in making waste less hazardous or non-hazardous at all. This includes developing mechanisms that are sustainable and applicable even at the household level.

Author

1 SOLID AND HAZARDOUS WASTES: -DEFINITION, SOURCES, WASTE MANAGEMENT AND CURRENT TRENDS IN HAZARDOUS WASTE MANAGEMENT

INTRODUCTION

Today, one of the many important challenges that humans haves been facing so far is the management of the various kinds of wastes that form during the production and usage of various kinds of products that humans use on a daily basis. Either they are the daily plastic wrappers on the foods or the ones that are collected during the production of medicines, various chemicals and the products that are used on an industrial basis. All such kinds of products and byproducts have some sort of negative effects on the ecosystem because of the fact that, they are not natural rather they involve certain synthetic materials which may not be able to merge in the environment in an easy way. Due to the various forms and the various compounds which form the waste materials of which are included in these materials, these waste materials can be categorized into various categories.

Most of the times we may categorize thewaste materials depending on the components it has and on the other hand it may be categorized on the basis of their effects on the environment or the ease of neutralizing the effects.

Basically the waste materials can be categorized as follows:

- Hazardous waste materials
- Non-hazardous waste materials

The waste materials which are categorized as hazardous, are mostly produced through the industrial and medical processes and materials. Due to the fact, when the various products are made on an industrial basis, there are a lot of chemicals and chemical processes which are involved and due to the various levels of processing the chemicals, there are many byproducts which are formed continuously.

Hazardous chemical wastes could be categorized further as follows:

- Industrial wastes
- Medical wastes
- Municipal wastes

Industrial wastes include the various kinds of byproducts that are formed due to the processing and production of various chemicals which can be in the solid form, gaseous form or in the liquid form.

Though we can say that, no matter which form of materials we have to deal with, there is a great need of having a proper system through which we might be able to manage the hazardous components that are there in the byproducts and the wasted materials which are of no use to the process or the industry that has created it.

Industrial wastes are considered the most hazardous kinds of wastes due to the fact there are numerous industries that are working to provide, synthetic and chemical products that include the process involving the combustion, burning and mixing of chemicals and releasing carbon dioxide, carbon monoxide and various harmful gases that not only create danger for the breathing organisms but it may disturb the whole ecosystem in a lifelong process that hurts the various stages of ecosystem and may affect the living beings severely at various stages of development as well.

Recently the main criteria that defines the hazardous industrial wastes includes that inherent properties including the chemical and physical characteristics like the toxic elements, the ignition power, the carcinogenic properties, and corrosion causing elements, and other such features making it harmful for the other substances around the materials that are being produced during the industrial processes.

Though if we look at the various elements, factors and the dealing systems in the developing countries we can see that they are not either defined properly, no categorized and not dealt with properly to make sure they would not affect the ecosystem to cause a long lasting damage as well.

Due to the fact when the developing countries are not taking much useful measures to deal with the hazardous wastes, there has been a number of issues that have arisen to hinder their way to manage the wastes properly.

Though, continuous development has been done to make sure that there should be a clear cut way and management system to deal with the classification and management and correct disposal of the wastes that would be beneficial for the environmentally friendly and would lead to recycle the wastes in a proper manner.

Further, the hazardous materials that are formed during the processing and production of the medicines and medical equipment and materials may also cause severe damages to the ecosystem if they are not handled with care. Due to many processes which are involved in the processing of medical materials including the antiseptic materials and other things, there might be a production of a large number of hazardous components including the various gaseous byproducts, liquids and the solid wastes.

In most cases, the waste materials are categorized solids, liquids and gaseous wastes and are further categorized as hazardous or nonhazardous and recyclable waste materials.

These waste materials are further treated as per the correct processing needs so that it does not affect the various environmental processes.

Municipal wastes also need proper handling as if it is not treated properly it can cause severe kinds of air pollutants and solid wastes which are not at all useful in any ways and may need to be disposed of properly.

As for example in China, the hazardous wastes that are produced in various forms either as industrial wastes or household wastes, they categorize them into household hazardous wastes and industrial wastes. In many developing and underdeveloped countries, only the industrial wastes and the medical wastes are treated properly.

In this way, they might neglect the household hazardous wastes that are produced due to the household wastes and due to the fact these are not treated properly, they might give rise to serious waste management issues later on. This happens when there is a lack of management of the wastes in terms of their classification, their categorization as recyclable and non-recyclable wastes and the lack of proper treatment plants and sites for the underdeveloped countries.

WHAT ARE SOLID WASTES AND HOW THEY ARE FORMED?

Solid wastes are those which are in the form of wastes objects and that are not usable in any way. They might further be categorized as recyclable or non-recyclable wastes. But it might happen when the waste is treated properly after classification and categorization on the basis of its physical and chemical properties.

There are mainly three classes and types of the solid wastes:

- Biomedical wastes
- Household solid waste
- Industrial solid wastes

Household Wastes

Household wastes or municipal wastes are the solid wastes which are produced by daily routine usage of various things like they could be construction and reconstruction wastes and materials, sanitation wastes, ansoalso, they could be wastes form the streets. They include the garbage and other disposed products that are thrown by the residential and commercial sectors.

Due to the increased urbanization and the usage of various packaged food products, there has been a great increase in the production of such solid wastes that has given rise to the challenges to manage the solid wastes that come from the household wasted materials and products.

It has been noted that as per information regarding the wasted solid objects in India during the year of 1947, the total waste production was 6 million tons while it was around 48 million tons in the year 1997.

And the main issue that is observable is that more than 25 percent of the solid wastes are not collected and treated properly.

Furthehethe70 percent of the solid waste that is being collected are not treated properly due to the lack of the facilities and disposal sites which are necessary to be there if there is a process that involves treating the solid wastes. Only 5 percent of the solid wastes might get proper treatment that is required.

The wastes could be the organic wastes like the kitchen wastes including fruits wastes or vegetables and other edibles. But due to the massive production of canned foods, the usage of cans, aluminum foils and plastics as well as many different kinds of non-degradable products that may cause an everlasting damage to the environment.

Further, the toxic wastes like old tablets, medicines and fertilizers, the batteries, polishes and varnishes sprays and pesticides are also included in the household wastes and that need special care when treating them as hazardous wastes because they might harm the surroundings severely.

Garbage: the four broad categories

Organic waste: kitchen waste, vegetables, flowers, leaves, fruits.

Toxic waste: old medicines, paints, chemicals, bulbs, spray cans, fertilizer and pesticide containers, batteries, shoe polish.

Recyclable: paper, glass, metals, plastics.

Soiled: hospital waste such as cloth soiled with blood and other body fluids.

Source: http://edugreen.teri.res.in/explore/solwaste/types.htm

In most cases, people may not think that the waste materials that are used in different ways in the form of household wastes are hazardous or not. But it has been seen that most of the municipal wastes that are no biodegradable and are in the form of solid wastes could cause severe or partial environmental damage and that may lead to further issues that are hazardous in many ways.

There different kinds of solid wastes produced as municipality solid wastes and they take time as per their elements to get disintegrated and mixed back into the environment properly. Like the organic wastes that include kitchen wastes are not hazardous and may take a week or two to be disintegrated back into its natural elements.

The type of litter we generate and the approximate time it takes to degenerate	
Type of litter produced	**Time taken to degrade**
Organic waste such as vegetable and fruit peels, leftover foodstuff, etc.	a week or two.
Paper	10–30 days
Cotton cloth	2–5 months
Wood	10–15 years
Woolen items	1 year
Tin, aluminum, and other metal items such as cans	100–500 years
Plastic bags	one million years?
Glass bottles	Undetermined

Source: http://edugreen.teri.res.in/explore/solwaste/types.htm

The paper of other paper like products may take approximately 10 to 30 days to get disintegrated. The cotton cloth and other such materials take about 2 to 5 months of time to be completely considered as disintegrated and resolved into its natural elements.

The wood and wooden items take around 15 years whereas the woolen products take 1 year to get completely decomposed.

The hazardous materials that play a significant role and may affect the environment on a long terms basis involves the tin and aluminum and other metal based products that take about 100-500 years to get completely disintegrated and mixed back into the ecosystem and that involves a lot of time indeed.

Some materials like glass bottles, plastic bags may take around one million or an undetermined period of time to be resolved.

Industrial solid wastes

Ecological contamination is the real issue associated with fast industrialization, urbanization and rise in living standards of individuals. For developing countries, industrialization was necessary and still this action especially demands to manufacture self-reliant and in elevating country's economy. In any case, industrialization then again has also caused serious problems relating to environmental pollution. In this manner, wastes seem to be a side-effect of development. The nation like India can ill afford to lose them as sheer waste. Then again, with increasing interest for raw materials for industrial generation, the non-renewable resources are dwindling step by step. Along these lines, efforts are to be made for controlling contamination arising out of the disposal of wastes by conversion of these undesirable wastes into utilizable raw materials for various valuable uses. The problems relating to disposal of industrial solid waste are associated with absence of infrastructural facilities and carelessness of industries to take legitimate safeguards. The extensive and medium industries situated in distinguished (conforming) industrial areas still have some arrangements to dispose solid waste. Be that as it may, the issue persist with small scale industries. In number of cities and towns, small scale industries find it easy to dispose waste all over and it makes troublesome for nearby bodies to gather such waste however it is not their responsibility. In some cities, industrial, residential and business areas are blended and thus all waste gets intermingled. Consequently, it becomes necessary that the neighborhood bodies alongside State Pollution Control Board (SPCB) work out requisite strategy for organizing appropriate accumulation and disposal of industrial solid waste.

Administration of Industrial Solid Waste (ISW) is not the responsibility of neighborhood bodies. Industries generating solid waste need to oversee such waste without anyone else and are required to seek authorizations from respective State Pollution Control Boards (SPCBs) under pertinent rules. In any case, through joint efforts of SPCBs, nearby bodies and the industries, a mechanism could be developed for better administration.

The significant generators of industrial solid wastes are the warm power plants producing coal ash, the integrated Iron and Steel mills producing blast furnace slag and steel melting slag, non-ferrous industries like aluminum, zinc and copper producing red mud and tailings, sugar industries generating press mud, mash and paper industries producing lime and compost and united industries producing gypsum.

Assessment of industrial solid waste administration issue incredibly varies depending on the idea of the industry, their area and method of disposal of waste. Further, to arrive at a suitable solution for better administration of industrial solid waste, assessment of nature of waste created is also essential. Industries are required to gather and dispose of their waste at specific disposal sites and such accumulation, treatment and disposal is required to be observed by the concerned State Pollution Control Board (SPCB) or Pollution Control Committee (PCC) in Union Territory. The following problems are by and large experienced in cities and towns while dealing with industrial solid waste

- There are no specific disposal sites where industries can dispose their waste;
- Mostly, industries generating solid waste in city and town limits are of small scale nature and even don't seek consents of SPCBs/PCCs ;
- Industries are situated in non-conforming areas and as a result they cause water and air contamination problems besides disposing solid waste.
- Industrial estates situated in city limits don't have satisfactory facilities so that industries can sort out their accumulation, treatment and disposal of fluid and solid waste;
- There is no consistent interaction between urban neighborhood bodies and SPCBs/PCCs to arrangement such issues relating to treatment and disposal of waste and issuance of licenses in non-conforming areas.

Bio-Medical Waste

Industrial and medical solid wastes are more toxic and more problematic as compared to the other kinds of waste materials. These material s are either in the form of toxic solids which are disposed of in a raw form or they might also be separated into various forms and categories. These materials are corrosive, toxic and may produce various hazardous gases as well. Biomedical waste (BMW) might be considered as any solid, liquid or fluid waste material including its container and whatever other intermediate products created during short term and long care focusing on observational, diagnostic, helpful and rehabilitative services for a person suffering or suspected to suffer from disease or injury or during research pertaining to generation and testing of biologicals during inoculation of individuals. From add up to amount of waste produced by medicinal services activities almost 80-90% is general waste equivalent to domestic waste. This comes from the administrative and housekeeping functions of Hospital and laboratories. The balance 10-20% of waste is considered hazardous and/or infectious.

Biomedical waste (BMW) is created in hospitals, research institutions, human services teaching institutes, clinics, laboratories, blood donation centers, creature houses and veterinary institutes (Pasupathi, Sindhu, Ponnusha, & Ambika, 2011)

Hospital wastes could be in the form of formaldehyde and other materials like phenols and medicinal products and equipment that include mercury. These kinds of products are severely hazardous and may require special treatment before they can be disposed.

Biomedical waste, also known as infectious waste or medical waste is defined as waste created during the diagnosis, testing, treatment, research or generation of natural products for humans or animals. Biomedical waste includes syringes, live vaccines, laboratory samples, body parts, natural fluids and waste, sharp needles, cultures and lancets (Shah & Ganguli, 2001) Improper administration of waste created in medicinal services facilities causes an immediate wellbeing sway on the group, the human services workers and on the earth. The waste created in these institutions essentially consists of solids and fluid, which might be hazardous, infectious and non-infectious. It has been estimated that up to 85% to 90% of the waste produced in hospitals is non-infectious (free with anyone fluids, which is similar to domestic waste) (Rajor, Xaxa, Mehta, & Kunal, 2012). It is the remaining 10% to 20% of waste that is of concern because it is hazardous and infectious. Likewise, waste that is un-segregated and not treated in the correct way would cause environmental pollution affecting the strength of the group. From waste audits done at several hospitals by a couple of NGOs, landed at some figures, which would now be able to be used and extrapolated for the entire nation. These audits must be conducted simply after satisfactory training on waste segregation is given to social insurance institutions.

Legitimate handling, treatment and disposal of biomedical wastes are vital elements of medicinal services office infection control program. Revise technique will help secure human services workers, patients and the nearby group. On the off chance that they are appropriately designed and connected, waste administration can be a generally viable and a proficient consistence related practice. Legitimate gathering and segregation of biomedical waste are essential. At the same time, the amount of waste created is equally vital (Diaz & Savage, 2003). A lesser measure of biomedical waste means a lesser weight on waste disposal work, cost saving and a more effective waste disposal system. Subsequently, social insurance providers should always attempt to diminish the waste era in everyday work in the clinic or at the hospital. To secure the earth and group wellbeing, the Ministry of Environment and Forest has advised, "Biomedical waste (Management and Handling) Rules 1998/2000 under the Environment (Protection) Act, 1986 that propel all hospitals, clinics, nursing homes, slaughter houses and laboratories to ensure safe and naturally sound administration of waste delivered by them.

Safe and viable administration of waste is a legitimate necessity as well as a social responsibility. Absence of concern, inspiration, awareness and cost factor are some of the problems looked in the best possible hospital waste administration. Clearly, there is a requirement for instruction as to the hazards associated with dishonorable waste disposal. Absence of detachment to the idea of waste administration is a noteworthy stymie to the act of waste disposal. A viable correspondence strategy is imperative keeping in see the low awareness level among various classification of staff in the human services establishments regarding biomedical waste administration.

Biomedical waste is created in all ordinary medical units where treatment of (human or creature) patients is given, such as hospitals, clinics, dental offices, dialysis facilities, as well as scientific laboratories, blood donation centers, university laboratories (Mukesh, 2001). Human services waste refers to all materials, organic or non-natural, that is discarded in any social insurance office and is not intended for some other use.

There are several methods that have been successful in the treatment of infectious waste. The following are the methods that will show the treatment that might be accessible at your office. The methods are:

Autoclave

Autoclaves are defined as closed chambers applying both warmth and pressure, and sometimes steam, over some stretch of time to sterilize medical hardware. Autoclaves have been used for about a century to sterilize medical instruments for reuse. Autoclaves are used to destroy microorganisms that might be present in medical waste before disposal in a conventional landfill. Autoclaves can be used to process up to 90% of medical waste, and are easily scaled to address the issues of any medical association. Small ledge autoclaves are regularly used for sterilizing reusable medical instruments while huge autoclaves are used to treat huge volumes of medical waste. Steam sterilization is most viable with low-density material such as plastics, metal pans, bottles, and flasks (Hegde, Kulkarni, & Ajantha, 2007).

High-density polyethylene and polypropylene plastic should not be used in this process because they don't encourage steam infiltration to the waste load. Plastic bags should be put in an inflexible container before steam treatment to anticipate spillage and drain clogging. Bags should be opened and caps and stoppers should be loosened instantly before they are put in the steam sterilizer. Care should be taken to separate infectious wastes from different hazardous wastes. Infectious waste that contains noninfectious hazards should not be steam-sterilized (Chandra, 1999). Waste that contains hostile to neoplastic drugs, lethal chemicals, or chemicals that would be volatilized by steam should not be steam-sterilized.

Incineration

This is demonstrated in which there is increase temperature causes dry oxidation. To lessen natural and combustible waste to inorganic incombustible to decrease volume and weight that can't be delighted, reused or disposed in outer land fields. The drawbacks to incineration include the vast capital and operating costs for present day technologies. The benefit of incineration is no Pretreatment is asked for and suitable for low heating volume over 2000 Kcal/Kg for single chamber and 3500 Kcal/Kg for twofold chamber. The waste should be less moisture as less than 30% and furthermore combustible (Pathak, 1998).

Thermal inactivation

Thermal inactivation involves the treatment of waste with high temperatures to eliminate infectious agents. This strategy is usually used for vast volumes (Patan & Mathur, 2015). Fluid waste is gathered in vessel and warmed by warm exchangers or a steam coat surround the vessel.

The types of pathogens in the waste determine the temperature and span of treatment. After treatment, the contents can be discharged into the sanitary sewer in a way that complies with State, Federal, and nearby requirements. This strategy requires higher temperatures and longer treatment cycles than steam treatment.

Gas/vapor sterilization

Gas/vapor sterilization uses gaseous or vaporized chemicals as the sterilizing agents. Ethylene oxide is the most normally used specialist, yet should be used with alert since it is a suspected human carcinogen. Because ethylene oxide might be adsorbed on the surface of treated materials, the potential exists for laborer exposure when sterilized materials are dealt with (Patan & Mathur, 2015).

Concoction disinfection

Compound disinfection is the favored treatment for fluid infectious wastes. Consider the following: Type of microorganism, Degree of contamination, Amount of proteinaceous material present, Type of disinfectant, Contact time, Other important factors such as temperature, pH, mixing requirements, and the science of the microorganism. Extreme disposal of artificially treated waste should be in accordance with State and nearby requirements.

Classification of Biomedical Waste

Nearly 75-90% of the biomedical waste is nonhazardous and doesn't cause any harm like any other municipal waste (Ponka & Kaski, 1996).

The remaining 10-25% is hazardous and can be injurious to humans or animals and deleterious to the earth (Fig. 1)

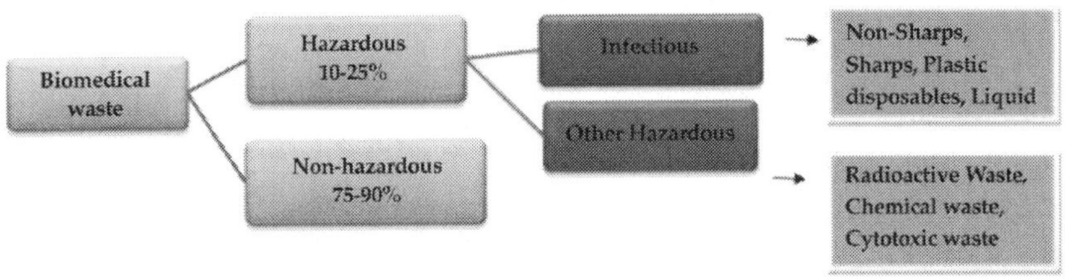

Figure 1: Classification of Biomedical Waste

It is critical to understand that if both these types are combined then the entire waste becomes hurtful. Aside from these the WHO classified medical waste into 8 categories such as General Waste, Pathological, Radioactive, Chemical, Infectious to possibly infectious waste, Sharps, Pharmaceuticals, Pressurized containers.

Wellbeing Hazards from Biomedical Waste

The disgraceful administration of bio-medical waste causes serious ecological problems in terms of air, water and earth pollution. The idea of pollutants can be classified into bio, chemical and radioactive. Land problems can arise because of the minor era of bio-medical waste and from the process of handling, treatment and disposal (Sharma & Chauhan, 2008).

Air Pollution can be caused in both indoors and outdoors. There are 3 types of Bio-Medical Waste that generates air pollution - Biological, Chemical and Radioactive. Indoor air pollutants like pathogens present in the waste can enter and remain noticeable all around in an institution for a long stretch in the type of spores or as pathogens itself. Synthetic Pollutants that cause outdoor air pollution have two noteworthy sources-open burning and incinerators. Open burning of biomedical waste is the most destructive practice and should be strictly kept away from Water Pollution is another real danger from Bio-medical waste. In the event that the waste is dumped in low-lying areas, or into lakes and water bodies, can cause severe water pollution. Water pollution can either be caused because of natural, chemicals or radioactive substances. The pathogens present in the waste can filter out and contaminate the ground water or surface water. Unsafe chemicals present in bio-medical waste such as substantial metals can also cause water pollution. Land Pollution is caused by the final disposal of all bio-medical waste. Indeed, even fluid effluent after treatment is spread ashore. Consequently, pollution caused to arrive is inevitable. Open dumping of bio-medical waste is the greatest cause for arrive pollution (Sharma & Mathur, 1989).

WHY THERE IS A NEED TO MANAGE VARIOUS KINDS OF SOLID WASTE?

In most cases, when it comes to managing the wastes and especially the solid wastes that are produced in the form of various substances in various societies, it is considered as one of the major necessities for any kind of community. Either you are living in a developed country or an underdeveloped country, there should be enough resources to manage wastes.

It is therefore a major thing to manage when there is a need to flourish in different fields for any country or community.

But it should also be noted that the management of any kind of wastes does not mean that it only involves the hectic work to manage all kinds of wastes rather the focus should be all the benefits that are involved when you are going to manage the wastes properly. So, if communities and management setups could focus on the various advantages and various facilities that are formed due to the proper management of wasted materials then it would be lot more interesting for the various governments to improve the various management working sites and definitely most of these setups would work more efficiently.

So it is better to throw some light on the positives of the waste management work that brings prosperity and may help in improving the overall conditions in the developing countries.

Here are some general benefits when there is a proper setup to manage the various kinds of solid wastes produced on an on industrial or domestic levels.

The health benefits

It is definitely a major benefit that can be observed in the form of the improved health and definitely a better community that is productive and active in many ways. When the wastes are removed properly and disposed using proper ways, it is quite obvious that the production of further hazardous wastes is reduced or even eliminated to provide a clean and healthy environment.

Most communities and countries that are able to control, minimize and reduce the various kinds of solid wastes are able to reduce the risks of health epidemics and that in turn supports a healthy community.it in turn also reduces the expenses of treating the epidemics and make sure that the developing countries would be able to progress and will be able to handle other related issues more effectively.

The chances of viral diseases and bacteria caused illnesses are also reduced and the people living there could enjoy a better health.

The environmental benefits

The environmental benefits are also there for sure. As if the solid wastes are reduced and managed properly, there is a greater chance that the environmental conditions would be improved. Solid wastes causes the solid and the water reservoirs to get contaminated and may reduce or disintegrate the beneficial properties of soil and the water in many ways.

By managing the various kinds of solid wastes the overall environmental conditions can be improved and it can reduce the risk that are caused to health as well as the various resources are also decreased. More trees can be grown and more water reservoirs can be maintained to support the ecosystem and a healthy lifestyle in the community.

The economic benefits

The economic benefits of managing the solid wastes are not hidden at all. Due to the fact when wastes are managed properly, there would lesser or no mess around the land and also there would be lesser chances of getting into health related epidemics.

The soil quality as well as the water reservoirs can be maintained properly and hence it can reduce the cost of living and maintain proper health reducing the cost of various measure that may take in a higher budget when government has to manage the health epidemics.

They can save a lot with this approach. Further, through recycling and reusing the various waste materials to create other products can also save a lot and create more options that are better for their use.

Using the recycled products can save a lot of money for the countries who are importing materiaormfrom the outside or other countries and may have to spend on the imports a lot. To counter such issues, reusing the same recycled products can save a lot of money and help improve the economic conditions of the country very comfortaby.

The community benefits

The community benefits could be noteworthy too. There could be many helpful possibilities that could help the community members to improve their overall condition by taking part in the waste disposal management work. It could be many things like there would be job opportunities for the various people who might have expertise in working in this field or there could be opportunities for those who are willing to take part in various processes and get more experience in such a wy.

THE CASE OF PENNSYLVAIA

As in the case of Pennsylvania, the solid waste management setup provides the following:

- The setup provides a number of jobs that may range from executive level job as well as junior level task and support the economic conditions of the local people.
- It serves to provide facilities for waste collection, disposal setups and recycling wastes for residential and commercial wastes so definitely contributing lot for the society.
- It support the whole process by purchasing lots of products made from other industries hence improving the condition for others as well.
- Provides lots of funds and donations as a part of improving the community and in turn the country level work.

Improvement opportunities for other industries

Due to the fact when the when the setups work for the management of the solid wastes, it needs lots of things, accessories and the various products to support the system. In this way, they have to purchase the products from the other industries and support their economic condition as well. For example, the Alliance Landfill has an expenditure of millions of dollars for a yearly budget and they purchase services and the various products to make sure their processes go on safely. They also purchase and get the various services from the various service providers.

NEED TO MANAGE HOUSEHOLD WASTES

Sometimes managing solid wastes is limited to managing the wastes that are produced by the medical industry and other industrial wastes and most of the authorities think or consider that these kinds of wastes are the only kinds of hazardous wastes that should be treated and nothing else counts as a hazardous waste.

While if the categories and the various kinds of waste materials are analyzed, it can surely be said that there are many different products and the various wastes that are produced as the waste materials in the domestic wastes and if analyzed there could be many things that might affect the environment severely and should be considered as hazardous.

These kinds of material include the glass wastes, the tin packs and the medicines that are wasted without any proper way.

So, it is better not to ignore the various hazardous or possibly hazardous waste that are formed and waste without any proper treatment through the household wastes or municipal wastes and may cause a long lasting damage to the environment.

The existing EU legislation does not give explicit rules to the administration (gathering and disposal) and definition/classification of household hazardous waste; consequently, the situation is fairly off-putting for the two citizens and waste authorities. There is no lawful prerequisite for householders to separate household hazardous waste at source, yet regardless of the possibility that it was the case, the implementation would be troublesome. Despite the absence of regulations concerning the specific accumulation and disposal of household hazardous waste in the EU, some countries give separate gathering schemes to certain types of household hazardous waste, oversaw at nearby level by municipalities, as in the UK, Netherlands and Sweden. In brief, the administration schemes are ordered as follows (Bernstad, Jansen, & Aspegren, 2011):

Created national systems as for instance in Austria, Belgium, Denmark, Finland, Germany, Luxembourg, Netherlands and Sweden

Nearby systems, as for instance in France, Hungary, Italy, Spain, Sweden and UK. In these cases there is a set number of neighborhood initiatives set up for separate gathering of household hazardous waste, usually a conveyance point at the focal point of enormous towns.

No composed system, in most eastern European countries where there is no separate accumulation of household hazardous waste yet mainly because of the absence of satisfactory infrastructure for waste gathering and waste recycling when all is said in done.

In the absence of obligatory regulations for the citizens, the gathering and, thus, the administration of household hazardous waste rely upon deliberate schemes based on the EPR and PS. All in all, citizens must sort and segregate the household hazardous waste, store it, and afterward transfer it to the curbside or more probable transport it to an offsite accumulation site. In spite of the fact that research has supported that curbside accumulation is the most helpful gathering system for households, offsite drop-off remains appealing to waste administration authorities. The reasons are that curbside accumulation is more expensive, tedious to design, execution and work, while special provisions would be necessary for curbside gathering for some household hazardous waste streams (Wagner, 2013). In essence, another, parallel accumulation and administration system is required, sorted out by the authorities, the producers or retailers. Contrasted with simple disposal or commingled recycling, just like single stream recycling, EPR and PS impose extra efforts on citizens regarding material segregation and drop-off and, thus, comfort is of central significance. One of the real problems recognized internationally is that drop-off centers are regularly situated in the suburban of cities and citizens need to drive so as to use the facilities. Several studies have shown that more household waste is separated at source if households are given easy access to such facilities (Bernstad, Jansen, & Aspegren, 2011).

Household hazardous waste accumulation systems are summarized, identifying the stakeholders as well as obligatory and volunteer steps.

Volunteer support of citizens is the cornerstone of any system, since separation at source is a necessary step for any successful system. The support of municipalities, producers and at a lesser degree retailers is a necessary precondition. Curbside and off-site accumulation, either in the type of changeless gathering sites or accumulation activities and at some point, because of high cost, way to-entryway gathering systems are used around the world. Reclaim programs with the investment of retailers have been turned out to be a successful system during the last years, actualized everywhere scale and for a long time in US.

In Germany household hazardous waste accumulation is a typical practice used. Indicatively, paints, oils, different chemicals, batteries can be gathered by means of two unique types of accumulation schemes; recycling centers and versatile recycling centers (vehicles) which routinely stop at certain places (Eunomia Research and Consulting Ltd, 2002). The most widely recognized method for collecting household hazardous waste in Ireland is through a portable gathering service offered by privately owned businesses that are paid by authorities. Additionally settled facilities are given at Civic Amenity Sites, since these turn out to be more typical (Eunomia Research and Consulting Ltd, 2002). Super Drecks Kescht fir Biirger is the name of the system for the accumulation of household hazardous waste in Luxembourg. There are distinctive gathering schemes including versatile vehicles, recycling centers, special accumulation campaigns, and way to-entryway gathering. Another gathering system is installed for refrigerators (SuperFreonsKescht) and for WEEE. An intensive information system is significant piece of these systems. The campaign SuperDrecksKescht fir Biirger (accumulation and treatment) is financed by the Ministry of Environment. A first separation of the hazardous waste takes put during the gathering. All waste is transported to a special logistic focus where the final separation is finished. Half of the materials are reused. Alternate hazardous waste is dealt with in hazardous waste treatment facilities (landfill or incinerator) (Eunomia Research and Consulting Ltd, 2002).

In France, the EPR occupies a prominent part in the variety of administrative mechanisms towards improving recycling execution. The EPR idea is widened to include extra waste fractions. Among other waste streams, it refers to household hazardous waste (paint, varnish, glues, acids and bases) for 43,000 tons for each year (EEA, 2013). The gathering rate for batteries and storage cells is scarcely over 30%. Household hazardous waste gathering is done in 3 unique fashions, through:

- multi-material reclaim centers,
- portable gathering trucks that do rounds according to a preadvertised schedule,
- reclaim containers positioned in various locations (supermarkets for batteries, drugstores, and so on.) (Eunomia Research and Consulting Ltd, 2002).

In the case of Sweden, hazardous waste is gathered at recycling stations. In some municipalities the service of a natural truck collecting household hazardous waste at specific locations is given. Be that as it may, numerous households have no standard system for storing household hazardous waste, thus being a powerless link in this fragile chain. Petrol stations have previously been used for collecting hazardous waste, however as petrol stations are increasingly restricted from urban areas, the requirement for an auto for transport renders them not any more helpful than recycling stations. Some municipalities have furnished households with ecological boxes, where household hazardous waste can be set out for simultaneous accumulation with residual waste (Eunomia Research and Consulting Ltd, 2002). Summarizing the distinctive ways that EU part states manage household hazardous waste, it can be stated that countries like Germany, Austria, Belgium and the Netherlands have accumulation rates exceeding 30%. In the dominant part of cases, the solids are blended with household waste and non-hazardous industrial waste. The liquids are regularly disposed of in the wastewater arrange. This dispersal of harmful substances has consequences for all the treatment chains and negative effect on the earth. It creates extra administration constraints in incineration units, pipe networks and landfills. It involves extra operating costs and requires specific investments. It has also negative effect on the agronomical sludge outlets from purging stations and composted fermentable waste, when it does not discharge directly into the natural condition (Veolia Environmental Services, 2007).

Taking a step away from Europe, in Western Australia the household hazardous waste accumulation program sorted out impermanent gathering days and operates changeless household hazardous waste storage facilities. USA states and Canadian provinces are increasingly accepting EPR and PS, especially for WEEE. In any case, households seem to lean toward the warehouse or drop-off centerrather than returining the waste back to the supplier or retailer, except for expired prescription. In USA the execution of one-day group gathering programs is set up since 1980. As of now, in the USA, EPR/PS laws had been enacted in 32 of the States (Wagner, 2013). In the Maine state WEEE administration program is based on a Products Stewardship scheme, with the dynamic cooperation of retailers. One-day accumulation events constitute the most well-known practice in USA. Specifically, in New York State two basic types of household hazardous waste gathering programs have been held; some municipalities conduct household hazardous waste accumulation on scheduled accumulation days, while different municipalities have allowed household hazardous waste accumulation and storage facilities, frequently alluded to as allowed facilities. In Japan, the Law for the Recycling of Specified Kinds of Home Appliances was enacted in 1998 and the retailers are receiving used appliances from consumers, which afterwards are reused by the manufacturers, with the consumers shouldering cost for recycling (MEJ, 2012).

Minimally detailed information on citizens' cooperation in household hazardous waste accumulation events/initiatives is accessible outside USA where historical information show a normal support rate of 1.4e6% (Niedzwiecki & Adams, 2007). When all is said in done, it has been demonstrated that citizens should know about the potential risks of household hazardous waste so as to influence them to take an interest all the more effectively in separate accumulation schemes, while such schemes should furnish the subject with the opportunity to convey the pertinent hazardous materials frequently, all the time in an advantageous manner and without having to drive for a long distance.

BASIC REASON DEVELOPING COUNTRIES ARE NOT DEALING WITH THE PROBLEM CORRECTLY, AND WHAT THEY NEED TO DO?

In various developed counties, there are lot of things that have contributed to the development of the various waste management tactics that in turn have given them the power to manage all kinds of wastes in a proper manner. There are many things that they have solved quickly and massively to contribute to their war against managing the hazardous substances that are caused due to the various processes involved in the production of daily use products.

The most important areas that have helped to improve the overall waste management work for the developed countries are as below:

The understanding of the waste materials

The most important thing that has contributed a lot in managing the waste properly is their understanding of the waste materials and its causes. Due to the fact, in the developing countries there is a lack of understanding of the fact that there are certain chemical and physical substances which might contribute in the development of further waste products in various forms. There is also a lack of various systems that may help in identifying the various hazardous wastes and understand what are the reason for whicrethey are produced and how they can be managed for the betterment of the environmen.

The classification and identification system for waste management

There is a lack of identification and classification system in the developing countries that could help a lot in managing the waste properly by identifying the various categories and knowing its components. Only after identifying the wastes it can be managed easily and the various processes can be controlled to figure out the best method to control all the various kinds of possible elements.

The various management systems to manage wastes

The developing countries lack the various management systems that are involved in the management and handling of the waste materials. Most of the developed countries have many management systems and they know how to handle the various processes and the various levels of element formation to make sure that the wastes are managed properly without creating further issues for the community and the environment as well.

Possible facilities for managing wastes

Developed countries have got the right kind and right level of facilities that may help them improve the overall process of waste management and due to which they can manage things better. Possible facilities and the availability of the resources might help the management workers to create more helpful system to make sure the waste would get managed and disposed propely.

THE LEGISLATION FOR WASTE MANAGEMENT

There should be a clear framework for the management of anything that has an importance for the locals and for the ecosystem as well. In this regard the Basel Convention held on the control of the various transboundary movements of hazardous wastes and their disposal is being considered as the most important and effective regulatory legislative framework so far for the better management of the various hazardous wastes materials that are there to affect the environment.

Though to make sure that the various developing countries including China have adopted various different kinds of legislative standards that are based on the various standards that may reflect the international rules and provide a framework based on the regulations according to the kind of issues the country has to face on a regular basis.

The most common causes of developing such systems and frame works could be:

To minimize the production of the waste materials during the duel cycles and the production of the various kinds of fuels.

The also guarantee to have a high quality reduction in the production of the waste materials.

On the basis of the various kinds of research based frame works it was clearly stated that in the developing countries the Basel convention has not been implemented in the correct manner and that is why the hazardous wastes from the various countries still are there and the problem that may arise in this manner might affect the overall management of the wastes in the developing countries. According to the latest possibilities that have given rise to the formation of the various rules and regulations in the developing as well as the developed counties that support the sustainable waste management system that not only support a better environment but also provide a helpful way to make sure that there would be lesser expensive to carry out the various process to manage wastes and there will be lesser wastes that are produced in various different forms.

The advancement of waste administration law has taken after various diverse paths around the globe. There are clearly various reasons for this; one reason considered here is the legal structure in the States. Specifically, the distinctive trajectories took after by customary law and civil law States is considered, as is the contrast amongst government and United States. The essential conclusion drawn is that waste administration law created in customary law systems tends to be planned and administered based on specific cases, with principles drawn from the "sum" of the specifics. This can be contrasted with civil law systems which will probably choose individual cases based on overarching principles. In this present reality where waste administration law is increasingly created from a principle based approach, the ordinary result is resistance from precedent-based law states. This resistance is contended to be, to some degree, because of various world views on arrangement making.

The present focus for administrative discourse surrounding the administration of city solid waste is one of resource productivity. Waste administration legislators see themselves as having progressed past simply managing metropolitan waste with a specific end goal to address human wellbeing and natural issues, and now focus after avoiding waste, and recovering resources from waste that is created. This transition has changed between States. The American experience of state-based approaches to waste legislation can be contrasted with the European supranational legislative coordination, just as precedent-based law "base up" approaches can be contrasted with civil law "top down" systems. This essay explores this variety in approach between the States, focusing specifically on the contrast between custom-based law and civil law systems, and differences in approach between federations against unitary States. The interaction of the European Union and its part States in the field of waste law is also considered, as is the advancement of international law in the field of waste administration.

National Waste Management Law

European Union

The first piece of EU law relating directly to waste administration was the 1975 EC Directive on Waste, 75/442/EEC. As with all directives, the Waste Directive established the outcomes to be accomplished by part States, however did not specify the measures to be adopted to accomplish these outcomes. The Wallonia Waste case in 1992 was a milestone case interpreting the Waste Directive. This case used the principle that "ecologic harm should as a need be rectified at source", introduced in the Single European Act in 1987, to upset an unhindered commerce precedent established in a similar case in the Interhuiles case in 1983. This underlines the EU inclination to support principle over precedent. **The significance o**f the judgment in the Wallonia Waste case was reflected in Directive 91/156 which changed the Waste Directive, amongst different things to arrange the decision of the Wallonia Waste case for confined waste disposal, yet confirming that waste recuperation operations are to be liberated by national boundaries. The EU has also expanded its legislative limit into various point by point aspects of waste administration. Of the order issues, the packaging and landfill directives are specifically compelling because they demonstrate salient points of the EU's novel way to deal with waste administration. The Packaging Directive mandates targets for the extent of packaging to be recuperated, imperatively including a greatest recuperation rate. The most extreme is legislated for with a specific end goal to stay away from showcase distortions because of one State subsidizing recuperation, and depressing prices for recouped materials to the disservice of different States.

The Landfill Directive is definite and prescriptive, with the main genuine discretion left in terms of how the required diminishment in biodegradable waste to landfill is to be accomplished. Based on the recitals, a great part of the justification for the Landfill Directive appears to rest on three elements: the need to keep away spending little, polluting landfills distorting the market by attracting waste to them at the expense of all around overseen landfills, lessen the unnatural weather change capability of landfills, and decrease the long haul risk presented via landfills and surrendered dump sites. The two directives endeavor to control showcase distortions, and both mirror a specific aspect of the European impression of the free market. In seeking to assure both unhindered commerce and a minimum level of ecological quality across the Community, instruments such as the Landfill Directive empower more liberated exchange than a system of neighborhood rules would.

Attempting to accomplish the principle of universal ecological quality with regards to a myriad of neighborhood rules would require consideration and comparison of the natural standards at the source and the destination. Instead, by establishing minimum standards for ecological insurance, the EU can be assured that the earth is secured at all potential destinations. Thus, exchange can happen on a "level playing field" and can be free, operating without the scrutiny of the EU. The advancement of the EU can be seen to mirror its increasing capability in the field. In a similar vein to Germany, the EU has adopted an increasingly unified and prescriptive way to deal with waste control, with failings regularly rectified by an increased assertion of focal specialist. Waste law has also been increasingly shaped by the principle driven approaches with which civil law systems give off an impression of being agreeable, yet which are anathema to custom-based law systems.

United States of America

The American involvement in waste administration demonstrates the strength of the states within its federal system, and shows its special origination of administration, the earth and their interaction. It also demonstrates the significance of the Courts in a custom-based law system. The law around waste administration in the US started to slowly crystallize with the suit between the State of New Jersey and the City of New York in the mid-1930s. In 1960s, the federal legislature chose to get involved in waste administration. Instead of centralizing waste planning, the federal Solid Waste Disposal Act 1965 accommodated "a national research and advancement program into enhanced methods of disposal", and a program of specialized and financial assistance to state and neighborhood governments. The genuine difficulty was to remain with the states.

This response was additionally strengthened in 1970 with the Resource Recovery Act 1970, which required an extensive investigation into resource recuperation by the official branch of government. The investigation requirements expected by the European approach of a federal expert developing a model of waste era, and all the more imperatively, controlling the era of waste through a series of tools. The official branch was slow in conducting this investigation, taking the stance that "the Federal Government has no business being in the junk business", and a federal strong position never truly created. Indeed, this disposition persisted for the following thirty years of federal legislation, however during this time the Supreme Courtwas has been exceptionally dynamic. In a series of judgments in the late 1970s and the mid 1990s, the Supreme Court routinely toppled attempts by the states to restrict the interstate development of waste because they presented barriers to interstate exchange. This may be contrasted with the position adopted by the ECJ in upholding restrictions on interstate movements of waste.

Whilst both the ECJ and the Supreme Court maintain that waste constitutes "goods", the ECJ considers that waste represents "goods" that might be directed in its development to ensure the earth, and principles such as the self-sufficiency and nearness principles wind up noticeably vital considerations. The Supreme Court has maintained a clear inclination for the maintenance of facilitated commerce between states, building upon a long legal custom of valuing the Union over the individual states. In the absence of radical Congress activity and, specifically, a Constitution that is just infrequently altered, this approach is unlikely to change. The EU, then again, encourages nearby distinction at the expense of integration, and has an exceedingly dynamic legislature with an explicit order within the EC Treaty to manage ecological matters. The EC arrangement is changed regularly and extensively, providing more scope for the legislature to shape the strategy field.

The US has chosen not to incorporate waste administration. This is to a great extent due to the exceptionally strong legitimate basis of the states and the customary law lawful convention. The previous suggests that, where obviously equal outcomes can be accomplished from state and federal measures, the state based measures will be favored. This indicates that the US, similarly to the UK, prefers a "base up", case-by-case decision making process to a "best down", principle driven process. Under such a system, state based decision making is likely to be closer to the individual circumstances that the federal government.

Australia

Australian waste law took after a similar trajectory to that of every single other jurisdiction. Initially waste was of minimal concern. Populace densities were low and the wellbeing impacts of poor waste administration were minimal. As populations increased, nearby laws were introduced to control the general wellbeing aspects of waste, and later the ecological aspects. Australian legislation in the field of waste administration remains to a great extent the jurisdiction of the states. The favored federal approach is a coregulatory approach of a deliberate assertion underpinned by an administrative system. You can gather further information in National Packaging Covenant and associated National Environment Protection (Used Packaging Materials) Measure.

The legislative approaches adapted within the diverse states reflects outstanding nuances. The Victorian Environment Protection Act 1970 is fixated on integrated natural administration, whereas New South Wales created sectoral legislation, including the Waste Disposal Act 1970. South Australia adopted an alternate strategy again, as remains the main state in Australia to have container deposit legislation with its Beverage Container Act 1975. A revision to the Act in 1986, intended to impose a higher deposit on single use glass bottles than refillable glass bottles deposits, realized the main waste related case in the High Court of Australia (HCA): Castlemaine Tooheys v. South Australia (1990). This can be contrasted with the US Supreme Court case Minnesota v. Clover Leaf Creamery Co. (1981), as the Courts reached opposite determinations on similar facts.

The HCA chose that the correction was unconstitutional based on a test to the legislature's basis. The HCA contended that the legislation was unlikely to accomplish its stated natural intent of vitality productivity, as single use bottles would really enhance the vitality effectiveness of South Australia since these bottles were manufactured outside of South Australia. With similar facts, the US Supreme Court (on offer from the Minnesota Supreme Court) reached the opposite determination to the HCA. It dismissed the Minnesota Supreme Court's initial ruling regarding vitality proficiency, arguing that lone the legislature can shape an opinion on that issue. Similarly, it chose that impacts on interstate exchange were "not 'clearly excessive' in light of the substantial state interest in promoting conservation of vitality and other natural resources and easing solid waste disposal problems". These conclusions mirror a hesitance for the judiciary to "second guess" the legislature, and may stem from the US system of checks and balances between the various branches of government.

To summarize, the Australian situation as to waste law is fundamentally the same as that of the US. This is unsurprising, as the two States share a federal structure, and both have a customary law past. Where differences are overseen, like in the choices relating to container deposit legislation, this can be attributable to the strong separation of powers doctrine in the US, and the contrasting parliamentary system of Australia.

International Law

There is restricted specific activity on waste administration in international law, with two key sources of international law being suggestive of the international waste administration law administration: international custom and treaties.

International Custom

International custom ranges in a spectrum from general principle to customary govern, with a customary rule being a strong type of a normal principle of law. State sovereignty is thought of as a customary administer, and is represented in Principle 2 of the Rio Declaration. This principle gives States the privilege to misuse their own particular natural resources gave they don't to cause transboundary ecological harm, thus impacting on another State's entitlement to abuse its own natural resources. According to principles of State sovereignty, a State can't just create waste as copiously as it wishes, however it can also dispose of its waste in whatever way it sees fit with the key proviso that the waste must remain within its land.

The principle of precaution activity is all for preventing harm to the earth, and reducing and controlling activities that may cause ecological damage. In the field of waste administration legislation, the principle of aversion has turned out to be increasingly profoundly ingrained in the thinking of regulators and arrangement makers. A pattern can be seen where regulation deals with increasingly abstract notions of "avoidance". This continuum starts with the extremely solid idea of preventing pollution at the point of waste disposal, and progresses towards preventing the pollution associated with the consumption of the goods which generates the waste.

The prudent principle has a strong and strengthening influence on international law, and in a separate opinion to the ICJ Gabcikovo-Nagymaros case, Judge Weeramantry attested that the preparatory principle is a constituent of the more extensive lawful idea of sustainable improvement which has overall acknowledgment. Notwithstanding this opinion, the conduct of various States suggests that it is not yet a customary lead, a position likely to be strongly influenced by a State's legitimate convention.

Treaties

The two essential multilateral treaties regarding the administration of waste are the London Convention 1972 and the Basel Convention 1989. The advancement of the London Convention demonstrates an interesting pattern in the improvement of international law. The Convention is based on the premise that dumping is allowed unless explicitly disallowed. A 1996 convention reverses the approach of the Convention, prohibiting the dumping of all waste at sea with the exception of a small gathering of wastes. These wastes are for the most part inert or produced from sea based activities such as fishing. The shift from dumping being allowed unless explicitly precluded, to denied unless explicitly allowed is an impression of the growing significance of the prudent principle. The weight of evidence is never again on demonstrating that a waste will have an effect, rather it is to demonstrate that the waste won't have an effect. Of the States considered in this essay, all have sanctioned the 1972 Treaty however neither the Netherlands nor the US have confirmed the 1996 Protocol (as toward the finish of June 2007).

The London Convention and its 1996 Protocol demonstrates one trajectory in the advancement of international waste administration law, a trajectory based upon safety measure. The Basel Convention continues with this principle, yet tightens it to a principle of anticipation. The Basel Convention as originally concurred and endorsed seeks to control the transboundary development of hazardous waste and 'other waste'. The Convention prohibits the shipment of hazardous waste to nonparties, as well as prohibiting shipments of waste to States that have denied the import of such waste. Where the shipment is not restricted, the Convention establishes a system of control. It is remarkable that, except for Antarctica, the Convention does not preclude a specific hazardous waste destinations.

The proposed revision to the Convention takes the Convention substantially even more away from this esteem neutral system of control. As opposed to leaving it for States to choose on the off chance that they would prefer not to import waste, the correction proposes to preclude the disposal of hazardous waste from created in developing States, and to phase out this fare where the waste is to be reused. Through this ban, the Basel Convention moves from a system of control to a system of anticipation. It is outstanding that the majority of the States whose waste legislation is considered above questioned the initial ban proposal. Throughout negotiations, and following the advancement of European unanimity in support of the ban, the opponents were lessened to Canada, Japan, the US and Australia, who continue to push for provisions for recycling.

The continued opposition of Australia and the US to the Convention is a demonstration of vital philosophical differences between these States and the EU. Australia and the US, as demonstrated in their case law, have a long history of upholding the principles of interstate organized commerce, including the exchange waste. The EU, then again, is increasingly for the prudent principle, is more OK with notions of restricting exchange where the goods exchanged are wastes, and sees the uniform security of nature in various States to be a genuine objeve.

WASTE HANDLING FACILITIES

The most important improvement and a step forward in the developed countries that is usually missing in the various underdeveloped or devries iscountries is the management of the various waste management sites that are there to help people categorize the various kinds of waste and out them into various categories on the basis of their chemical and physical properties.

These sites are maintained and run by the government in the developed countries and help in the formation of better categorized piles of wastes that can be further treated as recyclable and non-recyclable materials.

These could be of following kinds:

The civic sites in the cities

The civic sites are formed and maintained in cities and the communities where the people can manage and send the waste in the form of various categories like glass, aluminum foils and cans, paper products, organic wastes and other such materials. They are put into various separate waste bins to make sure they are further handled in a proper manner.

The transfer stations

The transfer station refers to a building or processing facility for the various kinds of waste management and it is meant for the temporary waste disposal work and make sure to provide enough facility to dispose of the waste properly. This is mostly the place or the places and facilities where the community waste disposal vehicles would be disposing of their collected waste to be summed up and then it is carried to further facilities so that it can be treated properly.

The wastes are then carried to the next phase to treat according to the properties into the landfills, and in the recycling units or may be disposed properly after being declared as the hazardous waste materials.

BRIEF INTRODUCTION OF VARIOUS WASTE TREATMENT TECHNOLOGIES TO-DAY

Due to the fact, there are many different kinds of wastes that are formed during the various processes, and usage of the various components during the daily routine tasks, they might have varying characteristics.

On the basis of such properties the classification of the waste material makes it possible to put them through the various technological processes and the various procedures to form reusable products and the products which are easily biodegradable and can be mixed back in the environment without disturbing the environment.

There are many ways to make sure that the waste materials get mixed within the ecosystem easily without any issues caused in the surroundings. For this purpose, they are disintegrated into smaller pieces making them easy to disintegrate and get mixed into the soil without further processing and mixing of other materials.

The Process of Composting

The process of composting is sued to disintegrate and help in the degradation of the wastes through natural degradation process using the fungi and fungus. Such a method is used for decomposition of organic wastes. The organic wastes are disintegrated using the natural materials and give the end product of humus that is used to be mixed with the soil and can be decomposed further to release the various components that are actually found in the environment.

For proper composting processing there are some essential components that may facilitate or initiate the process of composting by activating and facilitating the fungi and bacteria. These substances would be water, carbon dioxide, nitrogen and oxygen. These could be added in various ways like adding the water and mixing the organic waste to initiate the process quickly.

When the process of composting gets started, the microorganisms, bacteria, protozoa and rotifers makes sure that the wastes are decomposed properly and are not left untreated while giving it a final form of humus that has easy to disintegrate particles which are in no way harmful for the ecosystem.

Composting involves the three phase including the mesophilic phase, the thermophilic phase and the maturation phase. These stages offer a continuous increase in the temperature of the waste materials and the process get accelerated as the stages proceed.

Natural composting is not fast rather it works slowly and it can be increased with the help of adding some substances and these may accelerate the composting process indeed. Further, the pathogens are removed to manage the overall composition and characteristics and to make it the way it should be.

The main materials that can be composted easily include, the organic wastes, the animal manure and bedding and human wastes and sewage sludge and waste materials.

Industrial waste composting involves in-vessel composting, vermicomposting, aerated static pile process composting, windrow based composting and all these kinds of composting procedures are used in the developed countries which involve the various levels of waste materials.

Other kinds of composting procedures are composting toilets, black soldier fly larvae composting and other such technologies which involve the usage of various organic materials and natural processes to support the process and to facilitate the composting stages and make sure to exhibit a sustainable waste management process that truly act in a helpful manner.

The related technologies that support natural decomposition and handling of the waste materials can be the Aerobic digestion through which the organic wastes are converted into the bio gas where the residues become a part of the sludge and may be further decomposed and become a part of the decomposed components that become the part of the ecosystem in a natural way by using the processes that place no harm to the environment.

The various procedures and facilities include the landfill sites, recycling procedures and the windrow composting procedures to get rid of wastes and make the disintegrated products available for natural disintegration.

Incineration

Incineration is a thermal waste treatment technology that utilizes combustion to separate natural materials as well as substances. Waste material is changed over into incinerator base ash, flue gasses, particulates, and warmth, which can thusly be utilized to generate electric power. The flue gasses ought to be cleaned of toxins before they are scattered in the atmosphere while the base ash can be utilized as a part of cement blend or arranged in landfill. There are four incinerator plant designs. All have the benefit of decreasing the mass of the first waste by 80– 85% and the volume (officially compacted to some degree in dump trucks) by 95– 96%, contingent on composition and level of recovery of materials, for example, metals from the ash for recycling (Rambol, 2006). This implies while incineration does not totally supplant landfills, it lessens the important volume for disposal altogether.

It is noticed that incineration has robust benefits for the treatment of certain waste sorts in some regions, for example, clinical wastes and certain hazardous wastes where pathogens and poisons can be crushed by high temperatures. Illustrations incorporate synthetic multi-item plants with assorted poisonous or extremely harmful wastewater streams, which can't be directed to a customary wastewater treatment plant. Japan, Sweden and Denmark are pioneers in utilizing incineration for power/warm era. In 2005, waste incineration created 4.8% of the power utilization and 13.7% of the aggregate domestic warmth utilization in Denmark. Incineration in Europe has developed over most recent 10 years and is relied upon to develop in the UK. In Europe, a portion of the power generated from waste is esteemed to be from a Renewable Energy Source (RES)' and is along these lines eligible for tax credits if operated in private. Likewise, a few incinerators in Europe are geared up with waste recovery, allowing the reuse of ferrous and non-ferrous materials found in landfills (Mehdudia, 2009). In the USA, with the increase in the quantity of vast reasonable territorial landfills and, as of not long ago, the generally low cost of power, incinerators couldn't vie for the waste (fuel). Since 2004, exertion has been made to elevate incinerator limit because of: ozone depleting substance benefits, enhanced air contamination controls, and ash recycling.

Recycling

An American building organization takes note that a blend of recycling and treating the soil is 46 times preferred at diminishing nursery gasses over incineration generating power (AEA, 2001). Materials that can be reused in any case, for which no formal recycling exists in the Seychelles include: a few plastics (packs and so forth), paper, cardboard, wood waste, gadgets (computers, telephones, TVs, and so on) and batteries. The additional time and vitality that it takes to make an item, the more critical it is to reuse it. Furthermore, non-biodegradable waste (basically plastics and other oil and gas inferred items) are considered non-inexhaustible because of the time (a large number of years) it takes to make the oil and gas. At the national level, there are a few techniques which can be utilized to diminish the production of waste. These incorporate redesigns of bundling, empowering the utilization of negligible expendable material important to accomplish the coveted level of security and accommodation; expanding buyer attention to waste decrease issues; and the advancement of maker obligation regarding post-customer wastes. These objectives might be accomplished through authoritative activity and the production of market powers and economic forces. When materials are gathered some handling facilities are required to restore the gathered materials to a usable frame. In little countries, where it isn't economically justifiable to burn through cash on reused material handling plants, private enterprise can assume a fundamental part given that adequate material is gathered and markets for the material exist. Universal exchanging of recyclables ought to be investigated, to the degree close-by markets are distinguished. Alert must be taken however to guarantee the cleanliness and uprightness of gathered recyclables. The deliberate or incidental consideration of poisonous materials in sent out shipments of blended materials has been recorded, and speaks to a potential risk to transportation contracts. A viability examination of option reused paper items ought to be attempted to supplant the seedling compartments that were delivered. The seedling compartments were not effective on the grounds that they degraded with moisture and the clients want to reuse seedling plastic holders. A conceivable item for viability examination is the take-away holder. By and by, Styrofoam holders are broadly utilized yet hard to reuse or reuse and are accordingly arranged to the landfill. An equivalent reused paper item would need to withstand warmth and moisture for up to one hour and after that could be treated the soil. Notwithstanding, it has been noticed that Styrofoam requires next to no vitality to deliver and is warm safe.

There is potential for Star as well as Samlo to give appropriate space and holders to a recyclable stockpile including: plastics, paper, cardboard, wood waste, hardware and batteries. The partition and transport of these recyclables to the focal collection point (e.g. Star Providence) would at first be the obligation of individual family units, association's as well as schools as the limit with regards to accumulate notwithstanding domestic waste gathering is likely constrained. As metals and PEP plastics are as of now gathered these would not need to be incorporated at the Star site however attention with respect to the sites of metal and PEP plastic recycling would should be progressed. Knowledge with respect to the 3Rs: Reduce, Reuse and Recycle, which incorporates cleaning, isolating and transporting recyclables to the redeem focuses or Providence is required to guarantee defilement of recyclables is kept away from. For instance, paper must stay free of food items and general dirt on the off chance that it will be useful for recycling; and anything with the recycle triangle can be taken to a redeem place for recycling. This instruction will aid to diminish the solid waste to be overseen in the land fill, incinerator or elective and rouse thought of approaches to create economic benefits.

Landfill

Landfill is a professionally engineered despondency in low populace zone, for the last disposal of left over after all the past strides of integrated waste management. Waste is covered in that misery keeping in mind the end goal to maintain a strategic distance from any hydraulic association amongst trash and condition including air and water. Landfill is generally favored on the grounds that it has the most stretched out scope of capacities and is minimally costly strategy for waste disposal. In Malaysia 80% of the waste is discarded in landfill, this could cause major issue in not so distant future on the grounds that the present landfill sites are left with less limit while the new sites are still under development (Manaf, Samah, & Zukki, 2009). While similarly in developing nations landfill is the minimum attractive alternative while most alluring choice is WTE.

In 1977, structural and maintenance management principles were established for conclusive disposal sites. The gauges separated the last disposal sites into three classes - controlled landfill, inert landfill, and secluded landfill - and stipulated that waste must be securely arranged at landfill as per their properties. The semi-aerobic landfill structure is utilized as the structure for controlled landfill. Unsafe waste, for example, waste containing overwhelming metal and PCBs can possibly adversely influence health and the earth, and should be discarded at separated landfill sites. Non-unsafe waste that can possibly contaminate open water territories and underground water, or may influence the earth due to gas, odor as well as vermin must be discarded at controlled landfill sites. Plastic waste, elastic debris, metal debris, glass, earthenware production and blocks that can possibly cause environmental pollution can be discarded at inert landfill sites.

The semi-aerobic landfill structure was developed in a joint report by Fukuoka University and Fukuoka City. A leachate gathering channel is set up at the floor of the landfill to expel leachate from the landfill, with the goal that leachate won't remain where waste is saved. Common air is gotten from the open pit of the leachate gathering channel to the landfill layer, which advances aerobic decomposition of waste. This empowers early stabilization of waste, keeps the era of methane and nursery gasses, which make it viable technology in the prevention of an Earth-wide temperature boost. By and large, the air exhibit in an aerobic situation changes carbon in natural matter to carbon dioxide, nitrogen to nitrification denitrification, and sulfur to sulfur particle, which lessen the era of foul odor and flammable gas. Also, an aerobic situation increases the movement of microorganisms that decompose waste, quickening the stabilization procedure. Then again, in an anaerobic domain, where oxygen is absent, natural matter is changed into volatile natural acids, for example, acidic corrosive, and afterward become methane gas or carbon dioxide, and nitrogen and sulfur are changed to smelling salts, amine, hydrogen sulfide, and mercaptan, which emit foul odor and adversely influence the living condition of the encompassing zone.A

ALTERNATIVE WASTE TREATMENT TECHNOLOGIES THAT WOULD BE HELPFUL IN THE DEVELOPING COUNTRIE

Today, the emphasis is on the development of the various procedures for the waste management techniques that are sustainable and can work in a way that is beneficial for the ecosystem, for the mankind and all the living things in the world. Such systems are the facilitators of bio-disintegration so that they could give better products and can serve to provide better ways to dispose of the hazardous wastes that maotnot pose any kind of dangers to the ecosystem around.

Aerobic Disintegration

This involves the degradation of the wastes mostly the organic wastes to form alcohol and ethanol and the bioconversion of the biomass to form mixed alcohol based fuels. The bio-drying of the wastes also is a common practices through which the solid wastes of organic nature are given the right composition and help in the formation of recycles and usable products.

Gasification

The gasification of products which may form the various gasses that are mixed into the air in a way that does not cause any harm to the environment and may facilitate the production of better substances or reusable substances to help living things live better.

Land farming, plasma arc waste management, pyrolysis, mechanical treatment and heating processes, sewage treatment UASB and waste autoclave use are some of the most common ways to handle the solid wastes and other kinds of waste materials in the developed countries. In the developing countries these could be adapted as the most useful practices in case if sufficient facilities are provided in the proper manner

AADVANTAGES OF WASTE MANAGEMENT

Waste management and especially when it is related to the management procedures of hazardous wastes including the industrial wastes, medical wastes and the household hazardous wastes, is a necessary process that requires special attention on a large scale. It is therefore the responsibility of the whole management system as well as the community members to make sure that the hazardous wastes are not left unattended and these are processed and handled properly.

So, in the developing countries if the hazardous wastes are handled and tackled properly, there could be many benefits of having a proper hazardous waste management setup that may not only help in producing recycle materials but will also help in managing the various waste materials that may not be disintegrated properly.

Management of the waste products to reduce unnecessary products

Through proper management of the waste products, the overall number of unnecessary products are reduced to little so that there are lesser things to manage and it can save time and money as well as reduce the need for having lots of human resources that may help in managing the wastes. Timely processing and treatment of the hazardous waste can help protect the environment.

Reduction of harmful wastes ensures the protection of the ecosystem

By reducing the number of various harmful products there is a greater chance to make sure that the ecosystem will be kept safe from getting negative impacts. When the hazardous wastes are disposed properly the whole ecosystem can stay safe from the various kinds of hazardous consequences.

Less soil corrosion and reduced loss of beneficial products

Soil corrosion is one of the major effects that are observed in areas where there is a presence of corrosive substances released from the industrial wastes. In case if there is a proper system to manage the solid and liquid wastes that are produced in the processing and production phases of industrial components, the corrosion issues and the soil pollution can be managed easily and the risks can be reduced to minimum.

Lesser chances of health issues

Definitely when there are no wastes and cultures for the hazardous substances and microorganisms, the health facilities can be improved and the overall health of the community gets better by reducing the health epidemics as well.

Improved economic conditions

Improved economic conditions are also guaranteed when there is a reduced chance of getting into health issues on a community level. Further, when the products are decomposed correctly they might give the products that are necessary and may reduce the need of importing things that countries might have to export otherwise.

Production of various usable products through wastes

Production of various different kinds of products with the help of the various by-products are also possible that in-turn help the countries in managing the various other processes as well

DISADVANTAGES OF NOT MANAGING HAZARDOUS WASTES

Due to the very fact that there are very less opportunities that are available in the developing countries, there are lesser ways to reduce the hazardous wastes and that is why, many kinds of hazardous conditions that are caused due to the waste materials.

The most common issues and disadvantages of not managing the various waste materials which are hazardous and are not easily decomposed in any way naturally could be as follows:

Health hazards and epidemics

Health epidemic is one of the core issues that developing countries have to face and this is mostly caused due to the mishandling and non-handling of the hazardous wastes caused due to the industrial wastes in the form of gases, solids and liquids that are not properly disposed. These hazardous wastes get mixed into the water and in the air and due to this the chemical affect the plants, the soil, the animals and also the agricultural production in a way that alter the various elements in the soil and may also affect the production of various kinds of other products as well.

The economic crises

The economic crises in the developing countries is another example of the various disabilities that are there to impair the functionalities of the weak management components and may also affect the overall economic conditions as a whole.

Most of the times when you can see that the developed countries that have managed to fulfil their needs and have started to develop further resources, have got the resources to manage the various recycling techniques that provide them with the products that they can use in place of the things they don't have and save them a lot of money that could have got spent on imports. Whereas when the resources are not managed properly they get wasted and developing countries have to manage the continuously increasing needs of industries and household needs by importing expensive materials that could have been saved if implemented proper techniques for waste management and materials.

The damages done to the ecosystem

The damage done to the ecosystem is quite obvious and when its compared to the various levels of wastes that are produced during the industrial processes, the effects spread over the land, the water and the air as well. The ecosystem may get disturbed because of the fact there is no way to reduce the effects until and unless it is being managed properly at the right time and in the right way.

BASIC THINGS THAT ARE REQUIRED TO MANAGE HAZARDOUS WASTES

To manage the various kinds of hazardous wastes that are commonly produced in the developing countries, there are many different steps that can be taken to confront with the various challenges that are there to counter and get things on the right path for the sake of managing various kinds of wastes.

The basic things that would be required in the developing countries to make sure they will enable their systems to counter the various hazardous wastes and to allow them settle within the ecosystem are as below:

- The proper knowledge of hazardous and non-hazardous materials:
- Proper experience in managing the various kinds of hazardous waste
- A definite criteria for classification of wastes
- Presence of disposal sites
- Management system of hazardous wastes
- Proper techniques to dispose the hazardous wastes
- Rules and regulations to manage wastes

Due to the lack of sufficient understanding of the hazardous substances and the ignorance of the various materials that could give rise to the various hazardous wastes, there as an increased risk of developing solid hazardous wastes and also in the form of the liquids and gases that may cause a long lasting and harmful effect on the various levels of the ecosystem. Though, some of the substances are easy to identify as having the components which are harmful and not easy to disintegrate in a natural way. But some compounds are very complicated and may need proper analysis and understanding of the components and the chemical properties as well as the physical aspects of the hazardous wastes.

It is also necessary that the various kinds of materials that are commonly produced in a community should be understood individually rather than a generalized understanding of the various materials.

Further, proper experience in the management, identification and classification of the waste materials is also necessary. If there is a lack of proper procedures and knowledge about the various classification processes, there could be nothing that we can expect to get better in the developing countries.

In case if there are sufficient facilities provided for the sake of proper categorization of the waste materials, there is a better chance to manage the wastes and get it disposed properly. There should be training and education sessions to make sure the experts are well aware of what they have to diagnose and how they can help in identification and management of the various materials that are of no use otherwise.

In another word we can say that when there is a need to manage the waste products, then there should be the presence of a properly managed system with enough rules that govern the management process and there should be a clear goal through which the identification and processing of the materials is done carefully.

Having properly explained systems and management procedures facilitated by the experts who know how they would handle the wastes can work if proper waste disposal sites and facilities are provided to carry out the necessary functions and to make sure that the wastes are disposed properly without harming the ecosystem.

With proper techniques and waste management rules it is a lotreasier for the developing counties to make sure that their country will be able to manage the hazardous wastes and that there would be a lot more opportunities that may get into action for the sake of betterment of the various levels of economic systems.

2 THE DIFFERENT KINDS OF HAZARDOUS WASTES, IDENTIFICATION, CLASSIFICATION AND THEIR SOURCES IN DEVELOPING COUNTRIES

HAZARDOUS WASTE AND NON HAZARDOUS WASTE

Hazardous waste is the waste material in any of the three forms that is dangerous for the human health and also for the surrounding environment. Thus it should always be dealt with care when we are disposing it after its generation in factories, laboratories or from any such kind of a source. The source could be different at the generated material could also differ in form. It could be in solid, liquid or gaseous state and thus its disposal also needs different techniques according to its nature. For dealing with such a kind of situation we first of all classify these wastes into different categories in order to make its disposal much easier and also more appropriate as well. Proper disposal of hazardous waste is a must they cannot be left unattended. If they are left unattended they can cause havoc for the living beings of any kind whether human beings, animals or plants.

These are the waste that do not possess any hazardous characteristic and thus are not harmful for the environment and also for the animal life on the planet. They do not have the components that are listed by the government or the health department as hazardous.

Hazardous wastes are generated from different sources they could be industrialrialnon-industrial. Industrial wastes include the material that is generated by the industrial sector of the production of the specified goods. It is the duty of the industrial sector to dispose such waste in a proper way or as ordered by the health department of the government. The nuclear waste that is generated from the nuclear reactors or any such source is never put in the open air because of its reactive and penetrating nature and also the threats of the radioactive rays. These rays can also cause certain allergies and also can cause skin cancer or these sort of diseases.

They not only affect animals but plant as well. That is why there is a need to classify the hazardous wastes in order to determine that what sort of threats are there and in what extremity if the hazardous waste is not properly disposed. This classification is based on the structure, reactive nature and the concentration of toxic in the waste. From structure we mean that whether the liquid is in solid, liquid or gaseous form. If the waste reacts with the gases in the atmosphere or is ignitable even at moderate temperature it is put in special sealed containers. The ratio of harmful material in the substances is checked in the laboratories which gives the clear picture that whether the waste has the hazardous effect or the proportion is so meager that it is harmless for the environment.

CLASSIFICATION OF HAZARDOUS WASTES

There are two major sources of hazardous wastes one is industrial and the second one is non industrial. The industrial waste is the one that is coming from the factories, industrial sector, and weapon department and also from minidustrialnon-industrial one includes houses or the domestic industry and the hospitals. The hazardous wastes that are generated from the industrial sector are more dangerous than the domestic ones because they are generated in a huge band alsoalso, they include certain chemical that are being used in the industrial sector and also nuclear wastes if they are dealing with uranium or other such type of matrial.

On the other hand the domestic waste is not only generated in a lesser amount but it could also be controlled in a better way as personal consumption can be curtailed easily as it depends on our own use of a particular thing and if not used theyt effectaffect the economy of the state or the job or earning of a person. While decrease of the hazardous waste in the industrial sector may also most of the time will cause a decrease in the production level as well can effectaffect economy and the personal earning of a person.

Generally the hazardous wastes are divided into three categories. The first and the most primary one is the one that is radioactive. It is most dangerous and most difficult to dispose because radioactive wastes can cause life threatening diseases even to the one who is near to the source and the process continues until whole of the radioactive material completely vanishes into thin air. We can slow down the process but it cannot be stopped.

The radioactive material cannot be buried directly into the earth because the radioactive rays have the ability to penetrate even. For this purpose certain containers are made which do not allow radioactive waves to penetrate. These boxes are usually made of lead. This way is expensive but still is a must. The second category is that of the toxic material which could be in any form solid, liquid or gas. Smoke or the waste material produced in factories could be an example of such wastes. These wastes are dealt in several ways. Sometime recycling or the eradication of the toxic material is done and in the things like water can again made suitable for drinking or irrigation purposes.

Another is the flammable and the inflammable one. The hazardous wastes that are inflammable are disposed where there is not heat or any ignitable material of some sort. Also it is kept in mind that to what extent the hazardous wastes are dangerous to the environment. Another form of waste that is generated from the chemical or pharmaceutical industry is the acidic waste that cannot be drained into water reservoirs and also cannot be left open and unattended. Animal manure or these sorts of wastes are not included in this sort because they have no harmful effect on the environment and also can be buried under the ground which increases its fertility instead of causing any sort of harm to the environment. A waste is usually considered toxic if it has LD 50 because at this level most of the living organism die after digesting the toxic waste also in the gaseous form LC 50 is the toxicity level at which the animals usually die after 3-4 hours of its inhaling. Another way of classifying the hazardous wastes in most of the countries is as follows

Explosive nature of the waste is checked that whether it can cause any sort of explosion at a certain heat or after reacting with any other substance in the air.

- The level of toxicity at which it can become harmful for the living beings after being mixed in the water reservoirs or when allowed to mix with the oxygen supply.
- The third category of the wastes is those that destroy the human tissues and a apid pace.
- THe last one is the excess amount of metal present in the waste that is not suitable for the living beings. Metals likb, Ag, Cr etcetc. are usually considered as hazardous.

Table 1: Sources and Types of Solid Hazardous Wastes

Source	Typical waste generators	Types of solid wastes
Residential	Single and multifamily dwellings	Food wastes, paper, cardboard, plastics, textiles, leather, yard wastes, wood, glass, metals, ashes, special wastes (e.g., bulky items, consumer electronics, white goods, batteries, oil, tires), and household hazardous wastes.).
Industrial	Light and heavy manufacturing, fabrication, construction sites, power and chemical plants.	Housekeeping wastes, packaging, food wastes, construction and demolition materials, hazardous wastes, ashes, special wastes.
Commercial	Stores, hotels, restaurants, markets, office buildings, etc.	Paper, cardboard, plastics, wood, food wastes, glass, metals, special wastes, hazardous wastes.

Institutional	Schools, hospitals, prisons, government centers.	Same as commercial.
Construction and demolition	New construction sites, road repair, renovation sites, demolition of buildings	Wood, steel, concrete, dirt, etc.
Municipal services	Street cleaning, landscaping, parks, beaches, other recreational areas, water and wastewater treatment plants.	Street sweepings; landscape and tree trimmings; general wastes from parks, beaches, and other recreational areas; sludge.
Process (manufacturing, etc.)	Heavy and light manufacturing, refineries, chemical plants, power plants, mineral extraction and processing.	Industrial process wastes, scrap materials, off-specification products, slay, tailings.
Agriculture	Crops, orchards, vineyards, dairies, feedlots, farms.	Spoiled food wastes, agricultural wastes, hazardous wastes (e.g., pesticides).

Source: http://www.biologydiscussion.com/waste-management/hazardous-waste/hazardous-waste-meaning-sources-and-management/16753

Now we will discuss each of the class in detail.

Chemical wastes

These include salts, organic materials and also acids that usually mix in water reservoirs and also in the soil when they are generated. Also in the form of water droplets the also float in the air. They are quite harmful for the both the plants and also for the health of the animals and humans. These types of wastes are produced in the explosive factories or in the drug industries as well. Chemicals are also mixed in water by the wastes generated in the petroleum industry as well. These are quite harmful for the health of the living beings. In human beings they can cause certain diseases like cancer and diseases of lungs if inhaled. They can also act as a poison for the plants as well when mixed with water or soil. The contaminated plants again can harm the animals as well that eat or consume them in any other way.

Flamables

These types of wastes are the most difficult one to dispose as they are hard to contain because of their flammable nature. They could be in any of three form.ge.g. solid, liquid or gas. In gaeyhowever, they are in most deadly state because they are difficult to contain in this way. They could be pesticides or spray paints. Aerosol sprays are also included in the group. These wastes are ignitable even because of a little friction or if kept in extreme temperature for a little while. These materials are always kept isolated so that they if catch flame can never cause too much damage to any other product. It should also be kept in mind that this material should never be placed close to an explosive material or close to things like power supply, fuel tanks and gas burners because this thing can increase the severity of the accident if one occurs because of any negligence.

Radioactive

These types of materials always keep on generating rays that are harmful for the living tissues. The source never becomes inactive and thus is stored in the lead containers through which the rays don't penetrate. Substance like uranium and plutonium possess this sort of nature. This is used in nuclear power houses and also in the medical fields as well. The substances which are radioactive remain generating the radioactive rays until they reach theifehalf-life. The process could continue for a number of years and can be slowed but not stopped if the material is naturally radioactive.

The radioactive rays can even affect the future generation as well as the effect the genes as well and causes malfunctioning of different body organs as well. There effect can remain in the atmosphere even for decades and will continue even with the same intensity as well. The chain reaction continues until the material such as uranium is fully consumed in the air. These rays are that much penetrating that they can affect the person even if he is sitting in the next room far away from the production source. These rays can penetrate through our clothes, through the cracks of walls and also can be transferred to other person through our clothes whom we meet such as the children of the house when the touch our clothes.

Domestic Hazardous Wastes

In our houses now a day we use such equipment or materials that cause hazardous wastes on a large scale. Another thing that is emerging these days is the e waste which means the dumping of the computers. The developed countries usually send the waste of the computers to the developing countries as well it is in the form of donations sometime but again the developed countries get rid of their hazardous waste in a technical way.

Sometime certain things that are commonly used in the houses are also included in the list of the hazardous wastes like lead batteries, paint used in houses, oil from the automobiles and also the car batteries that have mercury as well are all included in the list of hazardous wastes. Even different types of sprays used in houses are most of the time not environment friendly. Thus lead batteries should be disposed in a proper way and also things like engine oil or paints should be disposed in a proper manner.

Wastes generated from hopspitals

The term medical waste means the hazardous waste that is generated from hospitals or any other such center or facility. It may contain hazardous wastes like chemicals, pathological wastes and operation tools. Even human blood most of the time is contaminated if it is from the body of an unhealthy person. On an average there is about 38 tons of medical waste that is generated by a single developing country every year. A large amount of it is disposed by the municipal department and this waste most of the time is in the form of solid waste. Waste like contaminated bandages or other instruments that are used during the surgeries are most of the time a cause of infection among the people. Sometime an infected tissue that is taken out of the body of a patient during surgery also effects the people near to it if they are not properly protected of if the tissue is not properly dised offof by the hospital staff. The operation tools and also the tissues or the drainage material of a diseased person should be disposed of properly in a prescribed manner. The government or the health department should also establish some rules in this regard as how to dispose this hazardous waste that is being generated by the hospitals and is left unattended most of the time.

Identification of Hazardous Wastes

There are several methods when it comes to the identification of the hazardous wastes as we know that we cannot name all the wastes as hazardous. Hazardous wastes can be identified by the following methods.

They usually possess a poisonous nature or a certain amount of toxic in it that could be harmful for the living beings. The most noticeable thing in this regard is that the amount of toxic usually does not cause sudden death or damage but the process is always a slow one.

The hazardous wastes are usually reactive in nature they can react very quickly with the surrounding substances and make them polluted or unfit for the environment.

Ignitibility is another identification of the hazardous wastes but this thing should be kept in mind that not all hazardous wastes are ignitable though most of them show this sort of nature.

In the developed countries it is a trend that now the government or any department established by the government for the benefit of the environment publishes a list in which all the hazardous wastes are mentioned in order to educate the people and also the industrial sector that which sort of chemicals they should dispose carefully as their disposal under unsuitable conditions is prohibited by the government.

Let us first discuss the list system that is usually followed these days for the identification of the hazardous wastes as it is the easiest one. We don't have to check the wastes all the time we can just match the generated wastes with the given list if it is there in the list than it is hazardous otherwise it is not. The list is arranged in alphabeal der F,KF, ,PK, P and the last one is U. These lists are separated according to the nature of the hazardous wastes as some wastes are more dangerous than the rest. The amount of toxic also can vary according to the generating sources.

In the F list the sources are not of a specific nature the generated source could be an industrial sector or any waste water from different production plants. The hazardous wastes of this source usually have chemicals like methylene chloride or toluene in it. They wastes generated should never have these chemicals more than 10 % in ratio otherwise they are on the list.

The next one is the K list which contains in it the list of hazardous wastes that are usually generated form industries like steel refining and petroleum. Some of the chemical industries also generate such sort of chemicals.

In the P and U list comes those wastes that are usually generated in the production of things like the pesticides or other such sort of chemicals that are produced on a large scale commercially for busis. HowevertherHowever, there is a difference between the two lists and the difference is that of the extent to which a certain toxic material is present in the waste that is generated by a specific source which is of course most of the time is an industry. The material which has a low amount of toxic in it is put in the P list while the one with the larger amount of toxic and which is more harmful for the environment is put on the U list because of its extremity.

Table 2: Chemicals Included In The P List

MATERIAL	CODE	UTILIZATION
Arsenic Trioxide	P012	It is used in chemotherapy
Epinephrine	P042	Used in emergency situation and for the surgery of the eye.
Nitroglycerin	P081	Used in the treatment of the angina
Phenylmercuric acetate	P092	It is used in contact lenses
Potassium Silver Cyanide	P099	Used to kill bacteria
Sodium Azide	P105	Serves as preservatives in labs

Source https://www.intechopen.com/books/management-of-hazardous-wastes/

Table 3: Chemicals Included In the U List

MATERIAL	CODE	UTILIZATION
Acetone	U002	It is used in the pharmaceutical industry.
Bromoform	U225	It is used to make the patient unconscious during surgery.
Chloroform	U044	As a pain releaser, or as a sedative
Creosote	U051	Used as an antiseptic
Cresols	U052	Cure infections
Cyclophosphamide	U058	Used in the process of chemotherapy
o-Dichlorobenzene	U070	Used to eradicate germs
m-Dichlorobenzene	U071	Used as an active germ killer
Ethyl Ether	U117	Cures infection
Ethylene Oxide	U115	Used to sterilize the operation tools.
Formaldehyde	U122	Used as a preservative.
Formic Acid	U123	Useful for the treatment of the heart muscles

Source: https://www.intechopen.com/books/management-of-hazardous-wastes/

Now we will discuss the points on the basis of which certain wastes are put in these lists or we can say that how it can be checked that a certain material is hazardous or not before putting in on the lis,KF,PK, P and U.

Corrosive nature

The first one is the corrosive nature that is checked if the waste is in the form of a liquid or a semi-liquid state. The corrosive nature means that the liquid harms any metal or material that is dipped in it. This is because of the high PH value of the liquid. The liquid with corrosive nature usually have PH value more than 2. Such liquids are also harmful for the drinking and irrigation purpose. Most of the industries use chemicals like caustic soda which is with high PH. Also some industries use hydrochloric acid as well for the cleaning of the metals.

Flammable Nature

A waste is usually considered hazardous if it readily catches fire after exposing it to a temperature higher than 60 degrees. This could be checked by just reading about the ingredients that are being used in the manufacturing of the product. These things are written on the back of the product's pack. The flammable materials are usually present in oil paints, adhesive materials like gum or glue and also things with a certain amount of polythene in them.

Reactive Nature

Hazardous wastes are almost all the time reactive in nature. Reactivity means that the generated substance readily reacts with the things that are present in the environment in excess like air, moisture in air, soil or other things like the skin of animals or certain gases present in the atmosphere. Also when sometime these wastes are dumped underground they also react with the underground water reservoirs as well. The problem with the materials that are reactive in nature is that they most of the time become more poisonous after reacting with some other gas or phenomena in the environment. So that's why it is tries to contain them at the initial stage of their generation so that we don't have to separate them after their release. This thing save not only a lot of budget but also a lot of time as well.

Amount of Toxic

The amount of toxic is also checked in labs after doing several tests on the generated materials. If the amount of toxic is of a harmless nature than it is not included in the list of the hazardous wastes. For this purpose a process called the leaching procedure is adopted to check the amount of toxic in the waste material. For this purpose some quantity of acetic acid is put in the liquid after that several tests are made to check the amount of hazardous chemicals if any present in the presented waste. Hazardous materials like mercury, chromium and silver are usually tested in this way. One thing should also be kept in mind that if we mix hazardous waste intordousnon-hazardous waste the resulted material will have hazardous characteristic thus this mixing never leads to any conclusion and thus the test or the experiment is of no use at all in any circumstance whatsoever.

SOURCES OF HAZARDOUS WASTES IN DEVELOPING COUNTRIES

Hazardous wastes exist in all the three basic forms which are solid, liquid and gas. Also there are other categories that are being introduced in the current era as well which include the radiation form from the radioactive source and also the semi liquid state of most of the hazardous wastes produced in the industry. The mud form of the waste that is generated after the constant drilling of the material. This category is produced in the process of mining. The hazardous wastes are all around us in the form of pesticides that are sprayed on the crops, motor oil, smoke from the vehicles as well as from the industrial sectors, mercury and other this sort of materials. Thus this thing is quite clear that the hazardous waste is being produced not only by the industrial sector but also by the agricultural sector, cities, medical labs, arms factories, hospitals and mines.

That is why the issue of the hazardous wastes has become a serious threat even on the international level. Every coming year we are facing more and more disastrous situation as there is millions of a ton of hazardous waste that is generated from the mentioned sources and the developed nations usually dump it in the underdeveloped countries where the situation is even poorer as they have no proper technical facilities regarding this issue. The problem is that the hazardous waste start effecting the environment in a rather slow way so that we are unaware of the severity of the problem until it is reached and we are left with no solution most of the time. The dangers of these wastes cannot be estimated and even if we make rules at the current stage the material that has been dumped in the past will still pose a threat to all our plans. These hazardous wastes most have the time have a longer life cycle that's why they do not decompose that easily by any sort of process and also the wastes that are radioactive can never be decomposed. Still the industries are poisoning the water and the air on a larger scale and they are allowed by the governments to do so especially in the underdeveloped countries.

It is estimated that about 65-70 % of the hazardous wastes are generated by the petrochemical industry and the rest of the 35 percent is generated from the industries related to metal refining. The problem is quite troublesome not only for the developing countries but also for the developed countries as well. In the underdeveloped countries this problem is of a more serious nature because they not only lack the resources that can be used to dump this e but alsoalso, they receive most of the hazardous waste from the developed countries as well because the developed countries for the disposal of the hazardous waste send it to the underdeveloped countries.

The underdeveloped countries also lack the technical support needed for the purpose. If the developing countries stop their own chemical industry which is generating more than 35 % of the hazardous waste their work force will become jobless and thus the question of joblessness will arise in its severe form also causing a decline in the industrial sector as well. The governments in this region of the world are finally taking steps to stop the situation from become worse. The previous rules related to the environmental conditions are being amended and also the industries not following the rules are fined heavily. The government is also providing financial assistance to those industries that cannot afford the proper disposal of the hazardous wastes. Still approximately 8 million tons of hazardous waste is generated annually by each developing country which is quite threatening and dangerous for the future of the whole mankind.

Proper awareness is needed on the ground level in these countries in each step. Not only in the industrial sector but also the smoke of the vehicles the waste from the hospitals and the domestic waste could be controlled in a better way. There should be random testing by the government departments for this purpose in order to show concern for this serious problem. In the developing countries the main source of the generation of the hazardous waste is the industrial sector. The governments of these countries have failed to control these units as they are most of the time run on fuel instead of electricity because electricity is not being produced on a large scale in the developing countries. This use of fossil fuels pollutes the atmosphere to a great extent increasing the amount of carbon in the air and also certain harmful chemicals as well. The second source of this hazardous waste generation is the weapon industry that is now also dealing with the nuclear power in most of the developing countries as well. Even the electricity is being produced by the nuclear energy. The waste that is being produced by these power houses is not being disposed properly because the countries lack the technical skill and also the finance in this regard. The units also that are operating are not checked properly after a proper period of time and because of their malfunctioning as well radioactivity remains a threat for the people who are working in the surroundings. Now non industrialnon-industrial source of hazardous wastes in the developing countries which are of the usual routine like the medical sector and also the smoke of the vehicles but the problem is that the developing countries are dealing with the severe form of each one of this aspect and they have no proper solution or plan to control the situation in the near future.

METHODS TO DEAL WITH HAZARDOUS WASTE IN DEVELOPING COUNTRIES

While most of the developing countries are under great threat some are taking possible precautionary measures to stop it from becoming worse. These steps include the buying of the new technology for the proper disposal of the hazardous wastes. Steps are being taken to reduce the generation of the hazardous waste as much as possible. Also proper laws are being made against those who are not fulfilling their duty in this regard. The use of the fuel is being reduced and also proper classification is being mad also so that harmful materials could be extracted out of the main source before its use. Waste production in most of the industries is being minimized by the use of effective techniques and also the new machines that are more efficient these days. The developing countries still are facing several threats because the developed countries are still using them as a final solution for the disposal of their own hazardous waste. The developing countries as they have no proper legislation on this matter are receiving these hazardous wastes without having any wledge about hawhow to get rid of them. Most of the companies are allowed to use different names for the waste that is generated and thus they escape from the hands of the government when the material is checked according to the list of hazardous wastes that has been published by the government because the alter the name of that waste according to their own choice, as one material usually has different names in different languages. Also another thing that is not checked in the developing countries is the proportion of the hazardous waste that is present in the generated stuff. This thing needs proper testing and lab analysis but this is not done in most of the cases and the permission of the disposal is given to the source of the generated wastes.

Now proper labs even in the developing countries are being made on a large scale by the government and the health department to check such an issue on the spot. This has reduced the danger of toxic waste in a huge way but still the problem is yet to be solved as a whole.

One thing that is still not completely understood by most is that when we are talking about how to deal with the hazardous material it does not mean the disposal of the material only. It includes the transportation of the waste in a systematic way and also the storage of different categories of wastes before their final disposal is also quite significant. The waste in the developing countries is still disposed by the unlicensed companies that lack equipment andalso they use the old methods like land fill.

Also in order to make their job easy these companies mix the hazardous wan the non hazardousnon-hazardous waste and use that in land fill. This is quite dangerous for the health of the worker who is fulfilling the task and unaware of the danger and also it is the cause of serious land pollution in the long run in the developing countries as they are wasting their agricultural land in this pattern. Now in the developing countries proper assistance is being given to the private companies as well in order to educate them about the landfill technique and also to teach them the new methods of disposal such as recycling, refining and reusing of the generated wastes. Proper rules are passed and in most places they are being strictly followed by the local bodies even.

Prevention activities can be seen on a large scale both by the government and also by the industrial sector that is basically generating the hazardous wastes. Also the transport of hazardous material from the developed countries should be banned. There should be proper facilities in which hazardous wastes could be kept before the treatment or disposal. All the developing areas should report to a central command about the amount of hazardous waste produced in that region. Hazardous waste issue should also be given priority on other issues. Due to a weak infrastructure the developing countries are generating about 15000 to 18000 tons of hazardous waste per annum and they don't have the facility to deal even with have of the current rate. As the economies are developing at a rapid pace and there is a fight for progress and prosperity that could only be achieved by a strong economy. For this purpose the industries are trying to increase the production level and the governments of the developing countries in order to boast the industries for a strong economy compromising on the standards of disposal of the hazardous wastes. Most of them again hide it in the way of land filling and compromise the environmental conditions. Even in the develop even the non hazardousnon-hazardous are not distributed into the sectble and non recyclablesnon-recyclables. There are no rules and regulations for this purpose.

Thus the hazardous waste generation is on the rise. No plans are made in this regard as it is supposed to be a problem that will be left for the future generation to solve. The developed countries have developed proper methods and facilities for the hazardous waste and also they are keeping a check on the generation of the hazardous waste which is something the developing countries have failed to do. Thus the developing countries are still struggling to handle and manage their sources of hazardous waste generation in an effective way.

3 HAZARDOUS WASTE MANAGEMENT AND PROBLEMS

It is always a huge problem to dispose of hazardous wastes because disposal in most of the cases is not only difficult but also a temporary escape. We cannot hide the fact that the materials that are hazardous in nature cannot be eradicated from the face of the earth and also they never decay or decompose on their own.

Thus it is never easy to get rid of them. Different ways are there to treat them and make something useful out of them or at least something that is not harmful for the nature. The waste that is about to be treated is put under special controlled condition because any sort link with the atmosphere can contaminate the whole atmosphere in no time, making the whole process a useless one. The material that is about to be disposed is kept in special storage places and also the waste is transported in a very careful manner to the disposal plants. The problem is more serious in the developing countries which most of the time transport the hazardous waste to the developed countries as a final solution. The not only lack the technology but also the skill. Transportation methodology is poor or proper attention is not given by the organizations or the government in this regard.

The hazardous waste is treated by companies that are not officially licensed by the government. They still use the old methods of disposal of the waste like land filling that is one of the oldest ways of getting rid of the wastes. The method of land filling that is still used in the developing countries is quite a risky one and that's why it is banned in most of the developed countries these days. It is dangerous for the health of the general public. The waste that is disposed by the method of the land fill increase the quantity of heavy metals in the environment which is dangerous for those who are working on those spots.

The developing countries are facing a lot of problems in the treatment or the disposal of the waste because they do not have proper plans which should be followed in order to get rid of these wastes that are a continuous threat to the life of the general public. No proper guideline is given to those who given the charge of the disposal unit. There is no policy related to or directly about the land fill system. It is most of the time thought that the waste should be send to the developed countries as they know better how to get rid of the material or how to use it again.

This is the main reason because of which the companies in the developing countries never focus on reducing the amount of hazardous waste during the production process because they know that the afterwards can send it to the companies that can solve their problem. One thing should be kept in mind that it is not always possible to dispose of the whole waste without doing giving any harm to the environment. Another trend that is becoming the fashion of the time in the developing countries is that of mixing the commercial waste with the domestic one. This again is not a proper solution to the problem. In this way we are only hiding the problem and nothing else. Also the government issues permits without proper verifications. The waste that is generated by the companies is most of the time not tested by the government authorities before providing the permit. Thus no one ever knows what is present in the waste and the environmental and health issues remain under threat all the time. No one follows any rules and regulations and check and balance is maintained.

As the government in the developing countries not take it as a serious issue thus the infrastructure is always a weak one regarding the problem of waste management or its final disposition. Which category is considered hazardous is never defined clearly by the concerned departments. Thus the problems show that in developing countries special strategies are needed for the disposal of the hazardous wastes so that solutions like land fill can be eradicated and new methods of recycling should be promoted that are not only beneficial but also better for the environment also.

Also the method of recycling is an economical way of treating the wastes because in this way the materials that are useable again are extracted from the waste not only decreasing the quantity of the waste material but also making it less harmful for the environment. Heavy metals most of the times are extracted at first that damage the land during the land fill. The main thing that is tried in case of the hazardous waste is the reuse ostmost of the materials that are present in the waste by the help of recycling. In this way most of the metals that are also expensive can be extracted again for the use which is beneficial economically as well as for the environment. The developing countries also lack the skill and also the technology needed for the disposal of the hazardous wastes that's why again they remain stick to the landfill system though it is not favorable but is adopted most of the time because the developing countries are mostly unaware of other techniques and also the lack the finance in most of the cases for the disposal of the waste. Used oil is the main threat in most of the cases for which the developing countries possess no infrastructure and thus they destroy their own soil and also makes the environment harmful for the animals as well.

In most of the cases the waste are discharged in the water sources that are not cleaned afterwards or that which never remain in motion continuously throughout the year thus this contaminated water spreads danger for the living beings of any sort present in the area. Proper technical skill should be provided in such cases and it should be given primary importance in all the cases. The developing countries lack a lot when it comes to the complete disposal solution of the hazardous wastes. The proper transportation facilities for this sort of material and also the way it can be recycled for the reuse. Also which metals or extracted material can be reused and in what way and also in what quantity.

It should also be kept in mind that the extraction cost should not be more than the actual worth of the extracted material. Disposal process should always be economical and environment friendly. Proper men with the scientific knowledge should be assigned for the task. Also the proper training or establishment of a proper department is needed in most of the cases that can keep an eye on the whole process in a systematic pattern.

INDUSTRIAL HAZARDOUS WASTE

It is the waste that is generated from the industries and are a huge continuous threat not only for the environment but also for the but also for the creatures living in that particular environment including those (humans) who are the cause of that contamination. There are different types or categories of industrial wastes for example are reactive in nature while other possesses toxic substances in it that can contaminate anything in which they are mixed. Some hazardous wastes are flammable in nature that can cause massive damage if they catch fire in extreme hot conditions.

Different industries are the cause of such waste generation like the paper industries that uses bleaching agents in a large quantity, different chemical and fertilizer industries, oil refineries, cement factories and the most threatening one are that which is using nuclear source for the power generation or for the production purpose. Power generation sources these days are the cause of hazardous waste production on a large scale because they are using materials like uranium for the production. Some developing countries generate 14000 to 15000 tons of waste every year which not only contaminate their environment but is also dangerous for the world in the coming years.

These countries still have no infrastructure in this regard that can plan the correct disposal. Such department if established can play a vital role in this regard. The main duty of such a department is to first of all inspect that form where and in what quantity the hazardous waste is being generated. The source could be an industry or a factory. These could be forced to use new methods of disposal as are approved by the world organizations currently. The main cause is also the competition of the production of more and more goods is another reason. Every country wants its industrial sector to increase the production rate by using any means possible without assessing the waste disposal factor.

Most of the industries even today mix the hazardous wae non hazardousnon-hazardous ones and there is no one top inspect the situation that is a threat to the environment. The major mean that is adopted is again the land filling technique. Also no budget is spent for separating non recyclablenon-recyclable wastes which increase the quantity of the waste material. The industrial sector still lacks proper regulations for the disposal of the wastes. Plans should be made by the government in order to minimize the waste production and also the disposal of the waste that is being produced by the industrial sector.

Treatment and elimination are the two words that should be kept in mind when dealing with the hazardous wastes because just getting rid of them is not the proper solution when we are dealing with the material that is life threatening in most of the cases. Policies should be made by the government in this regard and also the waste management sector should start adopting the new methods of disposal such as recycling instead of sticking with the old ones. Techniques should be adopted by the industrial sectors as well that can minimize the production of such wastes in the first place. Mixing ond non hazardousnon-hazardous wastes should be banned on international scale and it should be checked strictly. Those units which fail to follow should be fined heavily by the concerned departments.

MEDICAL WASTE

Medical wastes are also a threat though the quantity is not the same when compared with the industrial sector because things like mass production or a competitive environment lacks in this issue, which was a major cause in the issue of the hazardous wastes of the industrial sector. Such type of waste is generated from hospitals or health centers. These days different research institutes attached with the medical field are generating such type of wastes which though not in a large quantity but still are a threat to the environment. All the things that are related to the health care centers like blood, toxic chemicals that are used, the radioactive materials that are these days used in different treatment and research purposes in medical field are also quite life threatening if left without proper disposal.

All the surgical tools after use becomes contaminated and can cause diseases in a healthy person if not disposed or wasted properly. Pathological wastes disposal is another huge issue when it comes to the disposal of hazardous wastes in the field of medicine. The most threatening situation is that in most of the countries there is no statistical data present that can tell the amount of waste that is being generated from the medical department. This shows that the importance of this issue is still not understood by many in the world. The health department or the concerned ones cannot take any sort of step until they are told the correct statistical measure of the waste produced and after that the ratio of the different categories of wastes and how they should be disposed.

The problem with the medical wastes is that recycling in most of the cases is possible as recycling of the medical tools or such things will always carry a threat of contamination which is life threatening and is unadoptable that's why it is prohibited all over the world. On the average about 40 tons of medical waste is usually extracted from the hospitals which is usually disposed by the municipal department and no proper attention is given for the proper disposal of different categories of wastes. Also the man power that is present in the municipal department too have no awareness about these matters in most of the cases and disposal of all the waste material is done in the same way without considering the needs of separation of different categories that cannot be burned w along with the other wastes. Radioactive substance should never be disposed in such ways that are being used in the medical industry. The radioactive materials should be dealt with proper care and experts are needed for its disposa proper equipmentsequipment while dealing with it otherwise it can cause more harm than good.

HOUSEHOLD HAZARDOUS WASTE

In houses sometimes because of the use of certain materials and chemicals hazardous wastes are produced these days which most of the time is disposed with the garbage without any further attention given to the matter. This is another very serious case because there is no proper department for the job and also the general public is unaware of the dangers of these hazardous wastes. These sorts of wastes are produced during the use of varnishes, paints and aerosol sprays etc. The masses are also unaware of the reuse of several of these wastes like the chemicals used for house hold needs, lead batteries, lamps, metal anectronic equipmentsequipment. Even polythene bags can be reused in several ways. Thus this aspect also is of much importance and the instead of establishing a separate department the main need is to educate the masses in this regard. In this was the household waste can be sort by the people themselves and they can dispose it in different sections created by the government. This can decrease the negative impact on our health to large extent. The main thing which we should understand is that land filling or the heating process is effective for the disposal of wastes but these methods cannot be applied on all sorts of wastes.

Waste	Practice						
	Equipment availability	**Legal framework**	**Public awareness**	**Technology**	**Inventory**	**Separation of hazardous waste**	**Incentives or penalties**
HHW	–	Nascent or non-existent	Lack of capacity and awareness	Low technology incineration	Lack of information on HHW	Unfriendly environmental collection, treatment and disposal system	Low incentives or penalties
Medical waste	–	No specific policy	Lack of capacity and awareness	Low technology incineration	Lack of information on medical waste	Inadequate collection, treatment and disposal system	Low incentives or penalties
Industrial hazardous waste	–	No specific policy	Lack of capacity and awareness	Low technology incineration	Lack of information on industrial waste	Comingled with MSW	Low incentives or penalties

Source: https://www.intechopen.com/books/management-of-hazardous-wastes/

PREVENTION OF HAZARDOUS WASTE

Now coming to the prevention of these hazardous wastes which is the main concern of the present era as the amount of these wastes is increasing day by day. This is the thing that is most preferred one all over the world whenever the policies are made. Now everyone knows that the generation of these wastes cannot be stopped. Especially when it comes to the radioactive material its disposal is the most serious one of all. Special trained staff and a proper department with the proper tools is needed for the job. To control these wastes in a better way different techniques are being adopted by these teams in order to save the public from the deadly fangs of radioactivity.

Our production and utilization give vast quantities of waste. An essential component when dealing with eco-cycles is in this way sustainable waste management. Articles that circle in society contain extensive quantities of various materials. Many are energy-demanding to create and contain substances that exist in constrained quantities. It is accordingly important that we oversee joint assets in a long-term way to accomplish sustainable cycles in society. Many articles furthermore contain substances that are lethal or hazardous and ought not to be discharged into the earth. This requires eliminating the most hurtful substances and taking care of accurately those substances that keep on being utilized. There is an eco-cycle strategy in Sweden connected to the Swedish natural goals. The strategy is gone for achieving a society with non-toxic and resource- effective cycles. This incorporates prevention of waste, changed examples of utilization, more proficient production techniques and waste management with a more noteworthy concentration on recycling. The normal cycle strategy takes a gander at materials and items all through their lifecycle keeping in mind the end goal to give as total a photo as conceivable of their ecological effect.

Utilization of energy for an item likewise must be weighed into the assessment. Sweden thinks of it as fundamental for the volume of waste to diminish in the event that we are to approach sustainable management of waste. Makers should as of now assess an item's natural effect in a lifecycle point of view when it is manufactured. Design and material choice, and in addition vitality utilization in manufacturing and utilization must be considered. What's more, sustainable cycles must be accomplished if a more noteworthy extent of waste can be recycled and reused. This recovers the two materials and vitality both, while additionally decreasing the utilization of hazardous chemicals and ecological issues in waste management.

Twenty-eight EU Member States in addition to three EFTA nations, Iceland, Liechtenstein and Norway, are legitimately obliged to embrace a waste prevention program under the WFD. As a few nations have local instead of national waste prevention, the aggregate number of projects (36) is higher than the quantity of nations (31). This survey covers the 30 programs in 25 EU nations and Norway that were embraced before 2015 ended.

Of the 30 embraced programs, hazardous waste is inside the scope of 25. Of these, 17 programs gave explicit targets for the lessening of generated volumes as well as the decrease or end of hazardous substances in materials and items. 4 Prevention of hazardous waste Only four projects, in Bulgaria, Italy, Latvia and Sweden, have set quantitative prevention focuses for hazardous waste, and seven, in Austria, Bulgaria, Cyprus, France, Germany, Latvia and Spain, have included prevention indicators. An absence of explicit goals and scope in programs does not really imply that the nation or district does not prioritize hazardous waste, as it might be incorporated into more broad destinations or scope classes. The nonappearance of prevention targets is majorly connected to the grouping of hazardous waste and the absence of a hazardous waste baseline crosswise over projects, bringing about few characterized prevention indicators. Modification of the waste arrangement framework and endeavors to fix waste measurements additionally add to the issue. The majority of the above have implications for the viable checking and assessment of implemented measures.

Waste prevention targets connected to the utilization or decrease of unsafe substances are explicitly incorporated into 17 waste prevention programs. Most of the current projects incorporate qualitative parts of waste prevention, for instance, those that intend to diminish the substance of hurtful substances in materials and items.

- Bulgaria means to diminish the substance of destructive substances in materials and items.
- Denmark means to make it less demanding for customers to purchase items and administrations that require less assets, contain less issue substances and generate less waste. It additionally plans to help material organizations in decreasing ecological effects in the production phase and making it less demanding to reuse and reuse textiles, including by lessening the utilization of hazardous substances.
- Finland intends to decrease the utilization of certain hazardous chemicals and supplant them with less hazardous choices.
- Flanders bans/keeps the utilization of hazardous materials in new structures and recovers hazardous substances amid the destruction of structures and infrastructure. Amid the pre-demolition reviews, particular consideration is paid to the nearness of hazardous waste. Demolition observing guarantees that no hazardous substances are available in the trash from which aggregates will be reused. The reuse or recycling of building waste and aggregates containing hazardous substances is disallowed and last treatment is imposed.

- Germany has an operational objective of decreasing and substituting hazardous substances.
- Latvia is focused on decreasing the amount of hazardous substances utilized as a part of the production of materials and items.
- Lithuania is attempting to decrease the mass of hurtful substances in materials and items.
- Poland has set a goal identified with items and production with particular emphasis on restricting the utilization of destructive substances.
- Portugal plans to act continuously to lessen the nearness of hazardous substances in items, materials and waste.
- Slovakia plans to additionally decrease the era of hazardous waste by supporting extended producer responsibility (EPR) for different items and by help of the EU Eco-Management and Audit Scheme (EMAS) and the Environmental Management System (EMS) as indicated by STN EN ISO 14001(9).
- Spain puts emphasis on decreasing the toxicity of substances in items.
- Wales intends to decrease the substance of hurtful substances in materials and items.

A few projects incorporate both qualitative and quantitative goals, with the exception of those of Cyprus and the Czech Republic, which incorporate just quantitative ones (EEA, 2015).

- Cyprus incorporates the target of lessening the era of hazardous municipal waste.
- The Czech Republic is attempting to balance out volumes of hazardous waste and plans to lessen them in the coming years.
- Estonia has a strategic objective of avoiding and diminishing waste era and its toxicity.
- Ireland's objective is to decrease the utilization of hazardous substances and the era of hazardous waste.
- The Netherlands intends to acquaint pragmatic measures in order to enhance item outline — less material utilization, less unsafe substances, more reused material and longer item life

— as part of its round economy framework.

- Sweden plans to manage and move stakeholders with the goal that ecological objectives are met, less waste is generated and items are free of hazardous substances irrespective of how much the economy develops.

Austria's Waste Management Plan targets both quantitative and qualitative prevention. The goal of decreasing the volume of hazardous materials and items was authoritatively set in 2002 in the Strategy for Sustainable Development (BMLFUW, 2015). Particular substances that ought to be banned or evaded are not recorded or portrayed in any of the projects. Decreasing the adverse effects of generated waste on human health and the earth is likewise implicit but instead vague (EEA, 2015).

The Swedish view has been that the volume and hazardousness of waste must be influenced to a certain point by measures taken at the waste stage. Measures to diminish the hazardousness and volume of waste ought to be fundamentally taken as part of work on items and chemicals. An essential condition to be met to empower the dangers to be diminished at the waste stage is that the on-screen characters concerned know which substances can be hazardous to the earth and health and which of these hazardous substances are contained in the articles they manufacture, handle or purchase. The pointedly increased turnover of shopper merchandise with more extensive scopes of comparable items and items with a shorter life, are critical factors for instance in the issues related with electronic waste that contains hazardous compound substances. Attire is another case of articles with a short life, substantial quantities going to general waste management. The best method for decreasing chemical dangers is to make a move toward the start of a chain of production and handling, as every single later stage, including the waste stage, are influenced. The flow of data in production and the handling process chains is, be that as it may, regularly inadequate, and there are a few variables basic this. The chains of production and exchange for articles are regularly long and can go through a few unique nations.

There are exchanging organizations here that are not manufacturers of either chemicals or articles and just market and exchange articles. It can hence be troublesome for a buyer in Sweden to follow back and acquire answers to inquiries concerning production strategies in numerous provider joins. Work to enhance access to data about the substances in articles, for instance by presenting data necessities, consequently should be sought after at a global level. Sweden's endeavors to decrease the hazardousness of waste by restricting the utilization of substances of high concern prompts items that influence human health or the earth in a less unsafe manner all through their lifecycle. Another instrument is the item decision administer in the Environmental Code, which expresses that any individual who attempts or expects to embrace an action needs to abstain from utilizing substance items or biotechnical living beings that might be dreaded to posture dangers to human health or the earth, on the off chance that they can be supplanted by such items or life forms as might be thought to be less hazardous.

A similar prerequisite applies to merchandise that contain or are treated with a chemical item or biotechnical organism.

An expansive and developing waste stream comprises of waste from electrical and electronic items. The turnover of brand new items is quick. A portion of the items, for instance computers and TVs, become hazardous waste when they are disposed of, and control of the course this waste takes is frequently insufficient or hard to execute. A vital measure is that when assembling as little hazardous material as conceivable is utilized and also to make the items simple to dismantle. The RoHS Directive managing the utilization of specific substances in electrical and electronic items is a particularly critical instrument in this regard. Moreover, Sweden has found a way to lessen the hazardousness of waste through its endeavors to decrease the utilization of hazardous chemicals globally.

Now coming to the prevention of these hazardous wastes which is the main concern of the present era as the amount of these wastes is increasing day by day. This is the thing that is most preferred one all over the world whenever the policies are made. Now everyone knows that the generation of these wastes cannot be stopped. Especially when it comes to the radioactive material its disposal is the most serious one of all. Special trained staff and a proper department with the proper tools is needed for the job. To control these wastes in a better way different techniques are being adopted by these teams in order to save the public from the deadly fangs of radioactivity.

Waste Prevention Measures

Measures to avert hazardous waste are incorporated into a few waste prevention programs. The projects incorporate, in addition to other things, arranged measures, exercises, activities and strategy instruments focusing on hazardous waste. Numerous different measures don't determine which concrete waste streams are tended to and may likewise allude to hazardous waste, for example, in green open procurement (GPP) rules. Cases of particular hazardous waste prevention measures incorporate the following (EEA, 2015):

- Austria: institutionalization of the building visa, demonstrating the utilization of crude materials and contaminations, and gathering of center information in the focal building and apartment enroll.
- Brussels capital district: in the framework of the ecodynamic organization name, sorting out gatherings for trade of food methods, so organizations can gain from each other. The point is to develop mindfulness raising efforts for little and medium-sized ventures (SMEs) on legitimate prevention and management of hazardous waste.
- Denmark: partnership for the substitution of hurtful chemicals.
- England: support of the Sustainable Clothing Action Plan, which intends to enhance the sustainability of attire all through its lifecycle.
- Finland: when experts draw up the rules for ecological grants, the hurtful effects caused by chemicals amid the waste phase are to be incorporated.
- Germany: incorporating extra item bunches in existing name schemes, for example, the Blauer Engel (Blue Angel), an outstanding natural mark that as of now incorporates asset efficiency criteria.
- Greece: the progress of WEEE reuse in family units. The measure is part of the more extensive activity to encourage the reuse or potential repair of proper discarded items or of their segments, eminently using instructive, economic, logistic or

different measures, for example, support to or establishment of accredited repair and reuse focuses and organizes, particularly in densely populated areas.

- Ireland: usage of the reexamined National Hazardous Waste Management Plan. The arrangement sets out priority activities for stakeholders for hazardous waste prevention, diminishment of hazardousness of materials and substances, and naturally solid management of hazardous waste. Prevention activities are coordinated particularly towards the pharmaceutical, agriculture, healthcare, family unit, and distributing and printing sectors.
- Latvia: development and use of controls for the confinement of certain hazardous chemicals in electrical and electronic hardware by the Cabinet of Ministers.
- The Netherlands: start of innovative changes for textiles to lessen utilization of chemicals, making the textiles sector all the more ecologically well-disposed and expanding administration life.
- Norway: bolster by the Ministry of Climate and Environment for stricter worldwide directions on the utilization of chemicals in textiles.
- Poland: development of cleaner innovations.
- Scotland: the Scottish government has resolved to chip away at enhanced design, reuse and remanufacturing over a scope of sectors and ventures, including those generating hazardous waste. This will add to a more roundabout economy.
- Slovakia: advancement of the EMAS and EMS as indicated by ISO 14001 for hazardous waste.
- Spain: substitution of hurtful substances in production forms.
- Sweden: inside the EU, seeking to guarantee that natural viewpoints, for example, longer lifespan, capacity to be repaired and the substance of hazardous substances are considered in the design of new items.
- Wales: cooperation of the administration with industry, process efficiency specialists and Natural Resources Wales to better comprehend how much industry has upgraded its procedures,

including hazardous waste. It will likewise audit the regulator's part in observing the execution of allowed industry.

INNOVATIVE PROCESSES USED FOR TREATMENT OF HAZARDOUS WASTES IN DEVELOPING COUNTRIES

In developing countries the situation is quite a serious one. The reasons for this are many some say that the lack of funds is the main cause of the problem, for others the lack of technical support is the main thing and for some it is the competition that forces the developing countries to use materials that generate harmful wastes for the increase in the production level. They compromise their environment for their economy and they are unaware that they are losing in a big way by sacrificing their environment. These countries should use innovative techniques along with the careful disposal of these wastes. Proper infrastructure is also needed that can manage the whole thing in a better way. Technical knowledge should be provided to the ground level as well. A new method that is being adopted these days is the phytoremediation. This method is attached with the planting of more and more trees in order to support the environment to maintain its current state by aiding it with plants instead of recycling or disposing the hazardous wastes in a better way.

The method is cost effective or we can say that it is economical and much technical knowledge or any sort of skill is not required by the workers. This method is being used in the countries lik Sri Llanka, Pakistan, Bangladesh, India etc. The problem with this method is that it is not that much effective as the process is slow and also the plants are not suitable for the job all the time. For example they cannot sustain in a radioactive attack thus stopping it is impossible. In severe conditions the growths of the plants slow down and they also cannot proper in the places where the water is already contaminated.

They occupy a lot place which is not properly available in all the polluted places where the hazardous wastes are on the rise. We cannot use the industrial land for the growth of plants. This is possible in the household hazardous wastes most of the time thus this method is not mostly used all the time by the developing countries. If it is used it is confined to a limited region that is usually not struck badly by the over generation of the hazardous wastes.

BEST PRACTICES OF HAZARDOUS WASTE MANAGEMENT IN THE WORLD

There are some practices which are quite famous in the world for their effectiveness. First of all the main need is that the manufacturing factories and those who are generating hazardous wastes because of the use of certain materials should try to limit their use so that the generation of these hazardous wastes is decreased thus making the control much easier for the concerned departments. The government should also establish such policies that can make the factories and industries to limit the generation of the hazardous wastes. The following are some of the best methods that are adopted in the world for the prevention and disposal of the hazardous wastes. In many countries these days the used oil and other sort of burnable material is again used as a fuel in the industry or in the vehicles. In this way the waste is used up to the maximum. The governments also all over the world are now taking steps for this purpose and finally showing concern in the disposal of the hazardous waste management. The municipality should also we well equipped and aided with well trained workers who know completely about the final disposition of different wastes. In this way they can separate the different types of wastes and different methods could be adopted for their eradication. Alternatives are also being adopted such as physic chemical treatment is becoming a fashion in the current situation. Proper calculations are made in order to check the generation rate of the hazardous waste so that preventive measures could be taken according to it because statistical measures is always the first step before executing any sort of plan.

In the developing countries the governments and different organizations are trying to educate the people in this regard by seminars and advertisements but still the conditions are poor. About 40 % budget of the municipality is spent on the disposal of the wastes but the problem is still critical the efforts are not enough. Thus it is a challenge for the countries which have weak economies. They depend upon the foreign help both in financial and as well as technical field. The financial resources are not up to the mark to fight against such an evil even in the developed countries. The main objective in such circumstances should be to curtail the generation of hazardous waste first of all. The institutes that are created for the prevention should start right from the root level by educating the masses about the threats of the waste. Rapid urbanization is another major cause of the problem of waste management.

In such countries where urbanization is not controlled the heaps of wastes and garbage starts appearing on the streets even which is the worst situation. The people in the cities should also work like a single unit and should be aware that this is not the duty of the government or the municipality alone. They in the long run will suffer more than anyone else. Everyone should keep this thing in mind that with the use of nuclear fuel the problem is not a simple one anymore. It is the most complex problem or threat that the man kind is facing at present. The collection as well as the storage of the waste materials is a serious issue.

Hazardous wastes cannot be stored close to the living areas as they most of the time have deadly effects. Their treatment or the recycling plants are always on the places where there is no population of any sort. For growing cities there should be proper plans for the drainage of the wastes and also if that thing is not possible the municipal department should try to make the people agree with their plans of getting rid of the wastes. The governments should also make master plans for the drainage of the wastes of the cities and villages. Another strategy is the change in the behavior of the villagers in order to discourage them from migrating to the cities. According to a calculation about 80 % of the waste management problem arises because of the behavior of the people. If the organizations and the government departments make the people agree to the right way of disposing the waste materials this problem can be solved up to the maximum level.

Another way is to make the people or the industrial sector pay for the disposal of the hazardous waste which they generate. In this way the government does not have to invest in this regard to a great extent and also the industrial sector will try to control the generation of the hazardous wastes because the more they generate the more they had to pay for the disposal of the waste. Proper containers should be provided to the hospitals and residential areas in which different category of wastes are thrown separately. So that the sorting problem is solve and the mixing of the haza hazardousnon-hazardous wastes is finally stopped right at the root level. The container with the hazardous wastes should be sealed properly during their transportation. The governments should eradicate the root causes first instead of paying attention to the major causes. Decentralized policy is also another way of solving the problem. In this way extra budget can be arranged on the city or district level as well by the local bodies

DECENTRALIZED MANAGEMENT AND APPROPRIATE TECHNOLOGY

The hazardous wastes cause much damage to the environment if not dealt or managed in a proper way. The two methods which are being used these days instead of the land fill are compositing and the bio gas production method. These methods not only save us from the land fill method but also provide us a suitable resource instead of polluting the environment. Compost on the other hand is used instead of fertilizer to increase the fertility of the soil. Even in the lawns and backyards of the house compost is used for the farming purpose at a low scale thus serving in two ways. Firstly helps in getting rid of the useless material and secondly helping in improving the land for cultivation purpose. Bio gas another useful fuel is used in the houses is low carbon fuel that causes almost no pollution in the air as compared to other fuels like fire wood. This thing also saves time of the women in rural areas who spend most of their time in collecting fire wood in the developing countries.

This also helps to stop or decrease the deforestation process on a large scale. Thus production of fuel and decrease in waste is done at the same time by these methods. The problem that arises in these cases is the space or land that is available to us and also the quantity in which the waste is generated is also a serious matter. If it is in small quantity it is of no use because continuous bio gas production would not be possible and if it is in a large quantity still the problem would be there of getting rid of the extra waste that is not needed for the job. Sometime because of the low quantity of the waste the production cost exceeds the worth of the product that discourages the people in the use of these methods. If the hazardous waste is collected from different sources again the transportation cost will upset the budget. That's why instead of the transportation of the generated waste by the municipal steps should be taken to make plants on the spots in order to save the transportation costs. This step also needs an understanding between the local public and the authorities so that they can work in cooperation which is beneficial for both. This stresses the need on the decentralized process that save the extra budget of the government and also large scale bio gas tanks have failed to fulfill their purpose because of the continual maintenance problems that were faced during the process by the municipal. Thus these alternatives should be allowed to work on small scale instead of the large scale use.

The main thing in this regard that should be taught to the people using and producing bio gas in their houses is to first of all separate the waste before using it so that harmful materials that can cause contamination could be separated from the rest of the material, because it is a must in order to ensure the safety of the general public who do not possess complete knowledge about the misuse of the contaminated waste and its harmful effect on them as well as on the environment.

The composting process

In this process we take the help of the microorganisms and other insects for the final disposal of the wastes by the decomposition method. In this method usually two ways are used first is aerobic and the second one is the anaerobic method of decomposition of the waste materials. Aerobic decomposition is with the help of the organism that need a large amount of oxygen during the process and in case of low oxygen they stop the process or in simple words start dying at once. The other method anaerobic is opposite in nature because less or almost no supply of oxygen is needed for the job.

The organism start the decomposition process in the absence of oxygen and stop the function of oxygen is provided which is the exact opposite of the aerobic process. Usually aerobic method is preferred in the disposal process se the odourodor produced during the process is not with a certain pinch. Secondly, during the process larvae and the eggs of the flies are destroyed or killed completely. The method of anaerobic needs technical skill to handle and run the process. The equipment needed is also tough to use and the process is also hard top control if it speeds up at once that's why aerobic method is usually preferred for the disposal of the generated wastes. Aerobic method has following steps which are properly followed during the job.

First step is the degradation of the amino acids and also the carbohydrates present in the waste into simpler components. The microorganisms in this first step start their growth at a very fast rate. This rate not only increases the temperature but also kills the pathogens present. The process continues for about 20 to 30 days.

In the second step the temperature decreases and the insects like worms and mites help in the decomposition of the material provided. In this step the material appear to be like a soil mass.

The last step is to stabilize the decomposed material for the further use. Nitrates are formed in the material and it begins to stabilize and also its fertility is enhanced. The whole process is done in alm6 monthstime.

There are several things that should be kept in mind during the process for a good end product. First one is the ration of carbon and nitrogen present in the material that is being decomposed. The ratio is usually 31:1 because microorganism can digest carbon in a better way when compared with the amount of nitrogen digested by them. Then comes the question of the quantity of moisture present in the substance. The microorganisms need a high amount of moisture because they cannot dissolve the nutrients present in the waste unless they are not dissolved in water. Thus without moisture the whole process will stop at once. Moisture also does not mean wet conditions because the process also stops if the amount of moisture is increased from a certain optimum level.

When we are using the aerobic technique we should keep in mind that the supply of oxygen should also be there in sufficient quantity. This can be done by moving or turning the material in the container once in a while. PH level in the whole process should never exceed more than 8. Also it should be tried that the pieces of the waste should be small because small pieces degrade more easily and at a rapid pace as compared to the large ones. Also the degradation also depends on the things that are decomposed for example bones and meat degrades slowly.

Heap and windrow composting (passive aeration)

It is the cheapest method that is usually adopted for the disposal of hazardous wastes. In this method heaps of waste are made and are provided sufficient supply of oxygen for a few days. It should be put under the shades to save or protect it from the rain. This process could be done in a small lawn or in the back yard of your house as much space is not required for the process but the process needs a lot of physical exercise. Though it is watered at specific intervals and also in a specific quantity. A wall is made around it in order to save it from animather out sideoutside interference.

Box, bin and barrel composting (passive aeration)

This method is still in fashion especially in the developing countries. Boxes made of concrete are used for this purpose, though boxes are sometime made of plastic as well but the ones made of concrete are more beneficial in many ways. They are first of all economical and secondly they do not have a negative effect on the environment. Another advantage of the method is that it needs a very small amount of area as all the waste is dumped in the container which compiles it in the form of a big heap. There should be proper functioning of the drainage in the container and it should be covered properly in order to save or protect it from the extreme weather conditions like the sun and the rain. For drainage purpose small gaps or holes are made in the bottom of the container. In this process it is not necessary to turn the waste material again and again for the oxygen supply. It matures on its own in about 35 to 40 days depending upon the quantity or the layering of the waste material.

Takakura Home-method (THM)

This method was first of all developed by Koji Takakura and that's why it got the name Takakura home method. It is a technique that started in Japan first of all and became famous for the disposal of the hazardous wastes in no time because it was not only cheap but also the material required for this method is all that is available in our houses and thus no special things need to be constructed or bought from the markets. The main reason of its fame first of all was that it does not produy sort of odourodor which is quite beneficial if you are using your back yard for the process. Due to this quality this process is done inside the houses even to protect the waste material from the insects and all this sort of outside interference. A wooden box of any sort is needed for the job. Fine holes are made in the box for the ventilation purpose. In the bottom of the box a husk is put that can absorb all the extra moisture that is drained from the waste matox further morefurthermore is covered with a net in order to save it from the insects of any sort. The waste material is further cut into smaller pieces so that the process can speed up because in this way decomposition becomes easy and is done at a rapid pace.

4 INDUSTRIAL WASTE MANAGEMENT IN THE DEVELOPING COUNTRIES

INTRODUCTION

Managing or the disposal of the wastes today is as important as their consumption is. This holds true for both the industrial and the domestic waste. The managing of the waste has become a big problem in the developing countries, where improvement on a large scale is required for enhancing the standards such as the developed countries. The developed countries in this regard could serve as a benchmark for the developing countries in order to have a standardized and a safe waste management system. However, the developing countries need to produce a customized waste management system in order to meet the specific needs for the waste disposal as per the exact needs and requirements of their country. The use if recycling/ reusing of the products have become inevitable both in the developed and the developing countries. It is a far more crucial need in the developing countries as that recycling is not only an economical solution but is also is known to be very safe for the environment.

BACKGROUND OF THE HAZARDOUS WASTE MATERIALS AND THEIR HANDLING IN THE DEVELOPING COUNTRIES

Differentiation between hazardous and non-hazardous wastes is extremely critical, so that the necessary steps could be taken toward their proper disposal and keeping the environment safe and health for the living beings. Before, we jump on to some of the different types of the hazardous waste materials found in the third world countries and other related discussion under the topic, a brief background of the waste material generation and their understanding is pivotal to give us all a complete picture. In the most ideal scenario, the hazardous waste treatment should be done under the most controlled and regulated conditions. This process include the likes of possession, storage, handling, and transportation and then the ultimate disposal of the waste material. In the developing countries, however, the treatment done for the hazardous waste material is done in a relatively uncontrolled and unregulated manner. There are times when the hazardous wastes are used to be exported from the developed to the developing countries. It has also been an established fact that for the majority of the third world countries, the waste materials are processed usually in the unlicensed facilities by deploying the most unconventional techniques, such as that of landfilling. Worst, the treatment done is not usually satisfactory or up-to-the mark, and are known to have the low environmental standards coupled with low technologies as well. The waste stream remainder is processed using the incineration plants and oil recovery etc. A critical part of the toxic wastes is still being mixed with the non-toxic wastes and results in some really serious kind of environmental impacts, related to the persistent organic pollutants and the heavy metals in the leachate. All this leads to serious health risk to the living beings, including the municipal workers and the environment at large.

In present time, the majority of the developing countries are known not have the dedicated or proper hazardous landfill disposal facilities available. South Africa, for example, is a country that have the toxic landfill disposal facilities. In many third world countries, however, there are significant challenges in order to develop the required toxic waste material landfills, which include the likes of regulatory requirements, technically skilled manpower, social acceptance, financial resources, very limited technology and difficulties in provision and maintenance of such facilities. The landfill by all means is considered as the least desirable option for handling the waste material hierarchy. Further, it is also recognized that several types of the non-combustible or non-recoverable toxic waste materials, it will not be the option that would be considered.

The satisfactory level of the monitoring, maintenance and the overall infrastructure for the transportation of the toxic waste materials are not available in the third world countries. The indiscriminate approach towards the disposal of the waste materials (solid type), which include the likes of used oil and e-waste materials are serious threat to the environmental conditions and that include the threat to animals and soil contamination as well. The hazardous or the toxic waste material, for example, the wastewater coming from the healthcare vicinities are very commonly discharged into the storm channels and those are not cleaned periodically.

The monitoring operations of the environment and health associate themselves with the mentioned practices are not done as a result of shortage of health and safety specialists and the technically skilled manpower to do the needful. Apart from all the changes mentioned here for the effective hazardous waste material management, there are also some other known challenges such as the organizational, scientific and the engineering challenges which in the third world countries need to be taken care of on emergency basis. The most worrying part regarding the third world countries related to the effective toxic waste management is that the, much required recycling base in such countries is very weak or at some places, it is not existing at all.

TYPES OF THE HAZARDOUS WASTES

Hazardous waste materials could be of several different types and each of them, based upon their intensity poses different levels of risks to the environment and the health of living beings. There is a need to categorize them independently so that they are understood at the first place and then proper plans could be established for their handling, recycling and disposal, as applicable. The following are a few different types of the hazardous waste materials which are generally found in the developing countries of today:

- **Industrial Toxic Waste –** The industrial toxic waste is defined as that waste material which is produced by the industrial sector and raises serious concerns and threats for the environment and public safety. Such waste type is characterized in terms of the acute, extrinsic and chronic toxicity and also by inflammability, corrosiveness and reactivity etc. The major critical source of the industrial type wastes are chemical, mining, mechanical, pulp and the paper industries, wood manufacturing factories and the cement production mills etc. The crucial industrial type waste material to be taken care of include the consumed oil, oil polluted materials and the used solvent as well. If we take example of Lebanon's industries, which is a developing country, then according to a research, the industries in Lebanon generated hazardous wastes in the range of 3000 – 15,000 tons which is mainly because of the inefficient waste handling infrastructure and also the deficiency of the environmental management plans, which will certainly increase the environmental impacts which are linked with the industrial activities. Further, there is very limited data available on the industrial wastes' quantities and other associated stuff related to its management. However, as a result of the fast development of the economies, no strong environmental quality standards, the presence of the mass processing and manufacturing industries and the ever-growing quantity of the industrial toxic waste would increase endlessly. In the meantime, in the third world countries, there is a confirmation that one of the most critical part of the toxic waste material is still getting mixed up with the non-toxic waste material, which is mainly done due to the landfilling and also due to generating the serious kind of the environmental impacts related to the consistent organic pollutants and the heavy metals in the leachate (landfill). According to the scientific research, there is an observation that the waste materials are not actually detached in origin into the toxic and non-toxic waste materials. Moreover, the non-toxic types of the waste materials are also not parted in order to put themselves for recycling. As a matter of fact, the scientific research also has concluded that the developing countries are known not have the rules and regulations that particularly deals with the industrial type of the toxic waste materials. On top of that, the third world countries does not have any form of the adequate waste management plans for the hazardous materials and neither the authorized centers in order to treat, manage and finally eliminate the existence of any hazardous materials present anywhere in the environment. The detachment of the entire waste stream for the potential reuse of the waste material is not actually implemented that professionally in the third world countries. This actually gives the opinion that there is an emergency need for the sake of establishing the waste management policies that are geared in the direction of stimulating industries, societies and the commerce for managing the toxic waste materials in a very professional and substantial way and also encouraging the move from the traditional waste material management ways to the known and modern 'cradle-to-cradle' technique.

- **Medical Waste** – The medical waste material is that waste which is produced from the health related institutions such as the clinics, hospitals, research institutes and the health care centers etc. Such kind of the waste materials is known to have the heavy chemicals, toxic chemicals and several other types of substances, which ae radioactive or genotoxic, sharps such as the syringes, knives, needles, scalpels and broker glass etc., infectious waste material, blood products and the bulk human blood etc. Currently, in the third world countries, there is no means of having the data (statistical) on the amount of medical waste materials being generated from the health related facilities. In most of the cases, however, the estimated count of the medical type of the waste materials are actually extrapolated using the number of beds placed or the bed occupancy in the health related facilities. As an example, in the city of Dhaka, Bangladesh, according to research, nearly 37 tons of the medical waste is generated from different clinics, hospitals and similar other types of the healthcare facilities in the city. However, according to the scientific research, the secure disposal of the medical wastes in Dhaka and also in the majority of the third world countries, remains at a quite lower level, which is quite alarming. In the meantime, considerable part of the generated medical waste material is disposed of as the Municipal Solid Waste (MSW) or the same is discharged without any proper control or monitoring. On top of that, in most of the third world countries, the majority of the incineration type of institutions are not actually fully, operational or not even properly maintained. The deployment of the simple equipment and the primitive technologies that may eventually lead to pollution, posing serious threats to the environment and the health of living beings.

- **The household toxic waste –** There are plenty of jobs done in the household which lead to the generation of several types of hazardous waste materials, which is because of the products that could contain the toxic substances. According to a research, it is a confirmed thing that many of such products include the cleaners, varnishes, motor oil, car batteries and pesticides etc. which contains the harmful substances that are injurious to health and also to the environment. As a result of this, the used materials or the leftovers from those consumer products are called as the hazardous or the toxic waste materials. Other types of the hazardous home waste materials include the likes of e-waste. It is also estimated that the third world countries would discard nearly 400 to 700 million of the obsolete computer systems by the year 2030. The majority of the household waste material is sent to the developing countries in the region of Africa and Asia under the false charades as if they are some sort of charity or donations given by the developed countries of the world, especially USA and the countries from the Europe region. At the present time, in the third world countries, there is an observation that some of the local and the national level rules and regulations have been developed and/ or may be drafted for the household hazardous waste materials management. As per the lax enforcement and its monitoring operations, the valuable form of the toxic wastes, which include the likes of lead, wasted mineral oils, batteries, wasted mercurial lamps, wasted photographic chemicals and specific types of the electronic waste materials that is known to contain the Printed Circuit Boards (PCBs) and heavy metals etc. are not properly disposeffof. Therefore, such things lead to considerable level of negative impacts on the environment and as well as on the public health directly. On top of that, in the developing countries, some of the household hazardous waste materials is almost always taken under the management as the municipal solid waste. The informal type of the recyclers are actually the most dominant sector which is recovering and eventually recycling the valuable wastes. Most significantly, in the third world countries, the known safe methods used for the disposal of the waste materials include the likes of landfilling, incineration and several other types of the special treatment, (such as the sterilization and the microwave treatment techniques) are actually not that well-discovered. Therefore, the predictability enhances the pollution of the hazardous household waste materials and things remain in disarray and far away from the desired solution.

METHODS FOR HANDLING THE INDUSTRIAL WASTES

For a company that is known to produce the hazardous/ toxic industrial waste then it is pivotal to look for the most optimized method used for the industrial waste management. The most useful methods for the needful are known to first of all reduce, then reuse and finally recycle whenever and wherever possible. The methods used should also guarantee not to cause any kind of harm to the environment and the public. The following is a brief account of the top methods for handling the industrial wastes in the developing countries:

- **Segregation and Recycling method** –The majority of the waste which is produced by the shipping, production and the packaging needs is not either compostable or reusable, but that is recyclable as well. The very first step in the industrial waste management program is the identification process where the items that could be recycled are pointed out. The next step in this sequence is to setup the dumpsters or the recycling bins and into them, the wastes could be sorted according to their identification. The majority of the recycling centers could handle paper, glass and the plastic type of the recycling. The majority of them are also capable of handling the scrap metal type of recycling, electronic material recycling, food waste material recycling and the cardboard material recycling etc. The need is to first segregate and also properly separate the wastes from the hazardous waste materials, non-toxic (solid) type waste materials and also the compostable type of the waste materials.

- **Composting Method** – The composting process is known to turn the organic type of the waste materials into the fertilizer, which could then be used to nourish the plants. The majority of the food waste type material is capable of being composting and also any of the unsafe type of the organic food items can be set for the safe composting. You can compost the food waste material, newspaper, leaves, very fine pieces of the cardboard material, sawdust and the straw etc. Compost is then mixed in the soil in order to provide the nutrients and at the same time, to encourage their growth. Composting is considered as one of the most effective methods for the reusing and recycling the waste material.
- **Landfills method** – The landfills method is the most common used in the developed countries such as USA and is also getting popular in the developing countries as well. The rule of the landfill is to only utilize the waste in it which is non-recyclable, non-toxic and the non-compostable type. When the waste is set to be consumed by a landfill then it is first confined to a very small area, compacted where is applicable and finally, the waste material is then buried into the ground/ earth. When the waste is first decomposed then it is set to release the toxic gases which could then be converted into the form of natural gases which are used for the fuel and power. The landfills are considered as the most cost efficient solution and they are

ideally designed for the purpose of minimizing the harm that could potentially be done to the environment and the health of the living beings.

CHALLENGES IN HANDLING THE INDUSTRIAL WASTES

The prime issues or the challenges which are impacting the management of the toxic waste material in the developing countries include the points given below:

- Insufficiency in defining the rules, regulations, planning and the required instructions on multiple aspects related to the collection and the handling/ disposal of the waste material.
- Insufficient existence/ operations of the institutional capacities.
- Lack of preventive activities related to the waste material.
- Exporting the toxic materials from the developed countries to the developing countries.
- Inconsistency in the waste material policy directions or the incentives given by the current local authority or the landfill operators coming from the private sector
- Mixing up of the toxic waste with that of the commercial and/ or the domestic waste.
- All of the environmental authorities (regional) does not register the waste types produced in their operational permit that uses the different waste material classification codes.
- Inability to quantify the toxic waste properly, which is being generated in the records (reliable).
- Insufficient collection points for the toxic waste materials.
- Best Available technology (BAT) concept for its expectations are not usually met with the current environmental permits.

- Any information given regarding the rate of the waste material production corresponding with the waste generation (recorded) is normally missing. In the third world countries, the information system related to the waste material is not detailed enough to have all the required information pertaining to the concentration of the toxic substances.
- Shortage of compliance and enforcement with all the existing rules and regulations.
- Absence of the operational private sectors working for the prevention of the toxic. Hazardous wastes.
- Inadequacy of the national facilities (dedicated).
- Lack of dedicated waste material committees and managers and also the plans which could work for the monitoring of the toxic waste material handling practices.
- Shortage of the conformance with the top-notch practices of international level regarding the handling/ processing of the waste materials.
- Lack of capacity building training programs, guidance and awareness on the management of the sectoral toxic waste material management.
- Inadequacy of the specified policies which are dedicated to the effective management of the toxic household waste material.
- Inadequate or improper hazardous waste management infrastructure which could also be self-sufficient.

All the points that are discussed above, give an indication that the required integrated strategies for the toxic waste material recovery are inevitable to have in the third world countries, which possibly could decrease the disposal rate of such material in the communal landfills and in the recycling process of the valuable materials contained inside those wastes. It is for this reason that the introduction and application of the high technologies is imperative. Furthermore, the need is also there for having the state-of-the-art promoting recovery and the recycling points, backed with strong toxic waste management strategies, which ideally should be technically, socially, environmentally and economically feasible for the developing countries. The best practices in this regard could be seen as following:

- Maximizing reusability
- Recycling should be a process inculcated into the waste management system
- Recovery of the potential stuff, including precious metals
- Proper disposal of the waste material which is safe for the environment and the living being

BENEFITS OF PROPER HANDLING OF THE INDUSTRIAL WASTES

The industrial waste material requires to be dealt for the hazardous wastes with immediate effect and in the most desirable manner, in order to keep the hazards of the industrial wastes, away from the living organisms and the environment. The most prominent example of the industrial waste materials include the items such as the things that range from sawdust to acids and the scrap metal etc. The industrial waste management is actually the set of the tried and tested strategies which are deployed in order to handle the industrial waste removal and its disposal in a very cost-effective and time-efficient manner. While, such quality related factors are met in the process, it is also noted that the processes performed are also in compliance with the set rules and regulations of the industrial waste management. If there is any need to know the precise details about the handling of the industrial waste management then the local industrial waste management authorities could be approached, in order to know about the right process and procedure to be followed. The following are some of the known and the most useful benefits of the industrial waste management:

- **Better for the environment** – The poor form of the industrial waste management could potentially lead to the waste materials that get into and become the part of the local environment and then they could possibly lead to the causing of some irreparable damage. As an example, the chemical pollutants types that are generated from the manufacturing and the several other types of the identical processed could possibly penetrate inside the water and that eventually reaches into the bodies of every life of the marine. Numerous types of the industrial chemicals have got the potential to build up when they are ingested. This means that the higher concentrations that build up into the bodies of different types of animals are known to be higher on the food chain, including the humans as well (potentially). The proper form of the industrial waste management will help big time in deterring such type of the results and at the same time, also ensuring that the less toxic type of the industrial waste materials are actually recycled and that is something which has its own set of excellent benefits.

- **Protecting the human health** – The improper type of the industrial waste handling and then its disposal could possibly contribute towards the increased number of incidents related to the different types of the medical conditions, along with the human beings living in the local region. As an example, if there is a run-off belonging to the cattle ranch which then seeps into the local river then the human beings or the residents that live down the river on its either side or nearby, then they could fall possibly ill with the diseases such as the e-coli type of the infections which is found to be the most common one. The most effective type of the industrial waste management would only become more effective and practical, when there is a certainty of such type of results not to be appearing at all. At the same time, there should also be a reduction in the quantity of the produced industrial waste in the first place.
- **Legal Liability** – Different type of businesses have the responsibility of effectively managing and disposing the industrial waste material that they produce during their industrial operations. If the companies are not able to fulfil the set expectations regarding the handling of the industrial waste material produced then automatically, the companies would become liable to compensate for the damages that are caused through their obvious negligence. Even in those scenarios, where the businesses are ale to win the legal lawsite regarding their negligence in the effective handling

and disposal of the industrial waste material, then they would still be losing out on their legal fees and also the time that they had spent over the legal law suit in the urge of defending themselves in the court of law.

- **Waste Energy** – The waste energy is known to be a by-product of some of the known methods used for the waste incineration. A, we know that the incineration could result in very large amounts of the carbon dioxide to be released and also several other types of pollutants then there possible have been some advance techniques where which work on capturing the type of energy produced in the incineration method and then deploy it for the generation of electricity. The result obtained is a form of a complex technique of re-using the different types in order to lower down any need of some possible future waste. As, this is still being part of only a smaller footprint of a range of industrial factories, but still it is a fine start and large number of advantages could be gained in the future for the proper and the most effective handling of the hazardous industrial waste materials.

LATEST TRENDS IN THE INDUSTRIAL WASTE MANAGEMENT

For the past few years, there is an interesting time in the handling, disposal and the overall management of the hazardous waste material and more specifically, for the eco-movement at large, which are related very much with the needs of the third world countries as well. The latest trend now is of the eco-friendly type of urban housing and the cardboard type of bicycles etc. which is just to name a few. These mentioned future trends would practically become much realistic when the developing countries would start to recycle a bit too much from their different types of the waste materialr, However, there are still a lot of those third world countries who are having very difficult time to recycle the available waste material in their locality. For the past few years, a lot of development and innovation has been seen into the recycling efforts and also the sustainability ventures. In order to get a good idea that where and how the future of the waste material in the industrial waste management is heading then the following are some good known examples to look at:

- **Ban on the plastic material** – it is now an open secret that the majority of the plastic materials would take hundreds of years, if not thousands, in order to photodegrade and that is still an extremely poor thing to be in store for the environment anyway. Further, the plastic materials are also known to be extremely hazardous both for the wildlife and the ecosystem. It is for this reason that a lot of cities are now looking to address the issue of the plastic material generated within their vicinities. In particular, the Styrofoam, has been widely discussed throughout the years and the different countries, cities and the towns have now been inclined to resort bans on the food packaging that using the plastic materials or the polystyrene foam. While, it is a durable and a cost-effective solution for packaging, but at the same time, its light weight is also vulnerable to make it spread through the wind and it could also absorb the compounds, such as the like of styrene into the groundwater and the earth at large. Apart from the ban on the Styrofoam, the plastic bags (shopping) and also on the plastic bottles seems to be a trend which will soon be followed.
- **Biodegradable plastic material** – The market of the biodegradable resins (plastic) have been on an incline for a number of years and it is presently expected to hike by a percentage of 19 per year, by the end of year 2017. The plant derived resins polylactic acid, labelled 'PLA' will continue to be leading the campaign in order to introduce the resins (bio-based) into a number of different industries and the markets. On the other hand, there are some possible different types of applications as well such as to include the clothing, car parts and several other types of the electrical components, where there is still a concern about the labelling of the different plastics 'biodegradable'. Without any effective type of the municipal composting and the recycling in the place for breaking down the plant based materials then such plastics would not degrade at all. At the time, when the polylactic packaging (acid) is mixed, in particular with different other types of the plastics, during processing, then it could also possibly contaminate the complete batch of the plastic being recycled, eventually, making it all useless. The risky inclusion of the resins taken from the feedstock could only managed properly, in case we start the

adoption of the widespread systems which are capable enough to rightly compost the plastic material. If not then the expected risk would just mitigate the consumer guilt with no provision of any real solution being provided. The doubt thrives anymore just as the market continue to grow for such type of the plastic materials.

- **Shift from paper to digital** – It was in the year 2012, when the president of USA, Mr. barrack Obama signed the legislation which demands the EPA to shift to a totally digital record systems by the year 2015. This would enable the commercial businesses and the retailers to report their hazardous or the toxic waste material directly to the EPA body through the use of an 'e-manifest' and that would make the commercial and the industrial waste to track a much more efficient and a streamlined process. The immense need for the digital form of the record keeping has never been more than today, across multiple government institutions and a large number of industries, especially at a time, where even the departments such as the Department of Veteran Affairs has also witnessed huge chunks of the paperwork and that could put the disability claims into huge delay of several years. Considering the efficiency of the digital record system to produce significantly less amount of the waste material, so there is a high chance that in the future for both the developed and the developing countries, the different government and private institutions and several other business types would be more inclined towards the digital system which will completely replace the paper use in those industries.

- **3-D Printing** - 3-D printing has opened new horizons in the manufacturing domain that never seemed to be practical earlier on. Ranging from the commercial use to mass production and then further down to personal and private sector as well, things are looking very good in the future. To our astonishment, the 3-D technology might even be employed for building a complete house in a dar, However, it has been noticed that such technology would increase risk of our dependence over the plastic material even more. On the positive side, there are some who are able to find the grinded-up type of plastic material from the vicinity around the home. Apart from that, the things such as Legos and several other types of the plastic waste material are also used. This all could be a very fruitful option for the printing technology. As a researcher about such trend of the future, just think if a complete novel market is opened up just for the purpose of consuming the plastic waste material for 3-D printing. Multiple recycled type of the plastic materials are considered usually much cheaper per pound as compared anyway with the virgin plastic types. The 3-D printing has got immense amount of the applications but still it is needed from our side to fully ensure that the material used for the purpose should be sustainable to the maximum.

- **Mandatory Composting –** It is just 5 percent of the 26 million tons of the food waste material that avoided a landfill in the year 2012. This is an indication that there is still some multi million tons of the food, which is sitting right at the bottom of the landfill and that could have otherwise been turned back into some healthy compost waste material for the municipal or the personal use. It is for this reason that all the municipalities across the globe are looking to initiate programs for the purpose of composting of the organic material and there are a few which are even making it compulsory. It is not only the eco-titan (urban) San Francisco playing with such a kind of a legislation. The Rhode Island has actually initiated this discussion and now even the city of New York also did this when the active mayor of the city was Michael Bloomberg. At this point of time, there could only be hope that such an inclined interest in the composting would continue to grow.
- **Energy generated from the organic waste material –** California in USA is considered as that pivotal place where the potential eco-technology is channeled and also the anaerobic technology type is considered as no exception. Sacramento County with its 'Sacramento BioDigester' is known for accepting food and several other distinct types of the biodegradable waste material and then process it in order to turn that into sustainable bioenergy. This very benchmark set in for the anaerobic digestion through the forthcoming things as an indication, the Sacramento digester is considered to be extremely efficient and this one especially has

the potential to process nearly about some 100 tons of the organic waste material in a single day. As a researcher about the organic waste, just consider that how good the situation would be if there are at least one of such systems in every major part of the country in USA and also in the developing countries where the things are still very much out of shape in utilizing the organic waste to produce energy.

- **Sustainable Innovation** – In the Umea Institute of Design, there was a Swedish student, who worked on the designing of a concept of an ERO, which is actually a robot that could actually recycle all the buildings made from rebar and concrete material. This inspiring concept also helped its designer (Omer Haciomeroglu) win the International Design Excellence Award which was given by the Industrial Designers Society of America. While, this stays only as a sort of a conceptual design at this point of time and an incredibly ambitious one, but the fact that a complete building made from concrete could be recycled (theoretically) is considered to be a massive groundbreaking achievement. The sustainable possibilities are being totally redefined constantly through the inclusion of incredible innovations like this one and here also we can anticipaseeseeing same kind of revelations to evolve consistently at a very fast pace.
- **Recycling of the cigarettes** – Despite the fact that smoking cigarette is a gross habit, but the other known fact is that nearly 38 percent of the litter found on the road is either a cigarette or any other form of some tobacco waste material. It is really a disgusting and a universal issue that we all still have to deal and live with ir, However, just recently, with the Teracycle's Cigarette Waste brigade program (recycling), an organization, a business and even a person over the age of 21 years is eligible to collect the cigarette waste or any other form of the tobacco waste material found anywhere from the streets or roads and could then send it directly to the authorities of the Teracycle program. Once that is received from the people, business or the organizations then the paper and the tobacco gets composted down and then the filters (cellulose acetate) are recycled into the industrial type of the plastic products, just as you would see with the shipping pallets. Almost an

identical kind of a program was even launched by the Vancouver city (city-wide) just recently and that came with the all required help from the Teracycle. Just as more people got awareness about the cause and the need of the program that there is some reliable and confirm solution to handle this enormous amount of the waste stream, so there is every likelihood that several other municipalities, even from the developing countries would look to follow in the footsteps of such type of suits.

- **Enhanced corporate responsibility** – There is absolutely no big deal in making the hollow sort of promises lauding that 'corporate social responsibility' but in reality, an increase number of businesses and the companies are seeing that the actions actually speak much louder and clearer than the words themselves. The age of the aware public and the aware consumer is upon us to decide or judge and following that, the companies would themselves take the initiative of enhancing their waste material (self-generated) recycling efforts and also would become far more verbal, in general about its sustainability. Greenwashing type of technique nowadays is getting extremely difficult in terms of managing it. This is because of the reason that the public today is far more vigilant and at the same time, they are looking to pounce upon the illegal kind of the sustainability efforts. Apart from that, there are also several kinds of the upsides to the businesses as well, which are striving hard to become sustainable, such as having the enhanced supply line efficiency and considerable amount of decrease in the industrial waste material. There is a possibility to see a lot more of this in the future, just as multiple well-aware consumers would continue to increase the demand, that the companies or business that they would purchase their goods from then they should ideally be a lot more environmentally conscious and socially responsible.
- **The never-ending issues with the e-waste materials** – According to a research, nearly 49 million tons of the e-waste were produced all alone in the year 2012. This count was given by the Solving the E-Waste Problem (STEP) program. The USA, all alone is considered to produce nearly lionmillions of e-waste material in the year 2010. The problem lies where most of this toxic

waste material is sent to the developing countries for disposal but remain unrecycled in gigantic size, electronic mass graves. In this regard, both the EPA and the UN's Global Partnership on Waste Management is consistently looking to precisely track the e-waste material generation on the international scaer, However, the e-waste issue still remains as pervasive as ever. As the urge for the sake of managing all this toxic waste material continues and becomes almost impossible now to ignore, so there is a likelihood that some large discussion would develop on the international scale in order to counter this global issue, for both the developed and the developing countries.

Steps to manage the industrial wastes properly:

Almost all of the industries of today make use of the hazardous waste in their normal routine and this when is improperly handled then there could be a very serious threat to the environment and above all the human beings. Such materials could be found in the everyday use items which include the likes of, the electronic devices, fluorescent lamps, solvents, oil and certain types of batteries. The laws at many countries require such items to be handled and disposed with utmost care. Again, for the developing countries, the situation is bleak, particularly the threat posed from the industrial waste material in tons, which remains undisposeder, However, it remains the duty of the different industries and the businesses to take the forever responsibility regarding the proper handling of the hazardous waste material that their industry generates. Before we jump on to the steps required for the proper management of the industrial waste, we should first understand that what a generator is, as this term we will be using quite often in the text to follow. A generator in the waste material context could be a person, business, organization or an industry whose process or the acts leads to the production of the hazardous waste material. Included in this definition of the generators are also those businesses, which are serviced by those companies who are known to provide regularly with the pick-up used solvent and the clean solvent for the purpose of recycling. Then, there is also another type of home-based businesses which could also be generators. In order to find out that you are a generator or not, then you are required to first evaluate your waste. Here, we willocusfocused to keep our discussion close to the industries, which served here as a 'generator'. The following is a description of the different steps required for the management of the industrial waste material:

1. **Evaluating your waste material** – The industries first of all need to evaluate their waste and the hazardous waste could be of any of the following types:

    a. The waste produced is corrosive, ignitable, lethal, and reactive or may be an oxidizer. The container labels of the products used and the safety data sheets could be of extreme use in this regard.

    b. The waste that is produced fails the TCLP or the Toxic Characteristics Leaching Procedure, which is actually a kind of a laboratory test that finds out the level of toxicity in a substance.

    c. The waste material produced has the Polychlorinated Biphenyls (PCBs) with a concentration of 50 parts/ millions or even more.

Depending on the amount of the toxic waste material that is being generated, the generator or the industry could fall into one of the following categories:

- **Very Small Quantity Generator** – Produces waste material which is maximum 20 pounds or may be less for a single month, filling ½ a drum with the liquid.

- **Small Quantity Generator** – Waste material produced is in excess of 220 pounds per month, giving a maximum of 4 liquid drums.

- **Large Quantity Generator** – Waste material produced is in excess of 2200 pounds or even more in a single month that fills 4 drums with liquid or even more.

2. **Applying for an Identification Number** – In the developed cities, such as Minnesota, in the USA, all the businesses, including industries are supposed to get a license for the said activity, with just a few exceptions. At a first step in this regard, a hazardous

waste identification number is issues which is specific to your site. The number obtained actually replaces the identification number of the US EPA. After this, the second and final step in this regard is to country office of the hazardous waste in order to formally apply for the license of the hazardous waste generator.

3. **Labelling waste material properly and storing it** – The toxic waste generated from the industries is required to be stored in a robust and a leak-proof container. The container is required to be completely closed, other than, when the waste material is either added or removed from it. The different types of the waste materials should be stored in different containers and all the individual container should be labelled properly with the phrase, 'hazardous waste', along with the date waste and a clear description of the contents, stored inside the container. Further, the container should be placed over a waterproof kind of a surface, having good enough space (aisle) in order to allow for convenient weekly inspections.

4. **Proper transportation and the disposal of the waste material** – The generators on their part are responsible for the management of the toxic or the hazardous material forever. Therefore, in order to ensure that the toxic stuff is properly transported and disposed, the transporter should be chosen who fulfils the following critical requirements:

    a. Is issued with the identification number of the hazardous waste material.

    b. Is a legal entity having a valid license or a permit to offer the hazardous waste transportation services.

    c. Is facilitated with trained staff for handling the hazardous waste.

    d. Is facilitated with the required liability insurance.

    e. Is carrying all the required credentials within his hazardous waste material transportation vehicle.

    f. Is able to transport all the ready-to-dispose hazardous

waste material into a permitted facility for the purpose.

5. **Manifesting the hazardous waste shipments** – Manifest is actually a multipage, specialized document for shipping which must accompany (with a few exceptions) for the shipments (off-site) of the toxic waste materials. It is a cradle-to-grave record, showing that what exactly happens with the toxic waste material by the time it leaves the premises of the company and till the time, it reaches the actual disposal destination. It is the responsibility of the generators for preparing manifest which would essentially accompany the hazardous waste material when it leaves the premises of the industry. The transporters at their end, as part of a good practice could also prepare their very own manifest, as part of the service that they offer, but still the responsibility of the manifest will always remain with the generator and not the transporter.

6. **Planning for the emergency situations** – The planning for the emergencies could be done in the ways mentioned below:

    a. Maintaining the spill and also the required suitable emergency response equipment which is placed within an easily accessible area.

    b. Proper training of the staff should be done for catering the different types of the emergency situations and all the staff should be well versed with all the emergency response procedures which are as per the requirements of your industry's site.

    c. The Small Quantity generators (SQGs) or the Large Quantity Generators (LQGs) are actually subject for some further requirements to be met as well. For this purpose, they need to approach their local hazardous waste regulator for the required information.

7. **Having trained personnel** – The training of all the employees is essential, who by any means have any role in the handling, storage or management of the hazardous waste material. This is also a mandatory step in order to get compliant with the rules of hazardous waste materials. The personnel must be well aware

about every kind of the waste's hazards and also the apt safety procedures and other pivotal aspects of the compliance.

8. **Keeping or maintaining the hazardous waste material records properly** – Unless specified otherwise, the following is a concise list of all the records which must be maintained properly onsite and should be available for the purpose of inspection for a period of 3 years:
   a. Manifest record and the receipts for the recycling.
   b. The inspection logs that are maintained on the weekly basis.
   c. The applications and the renewal documentations for the licenses/ permits.
   d. The analytical and several other similar types of documents/ reports.
   e. The forms required for the Land Disposal Restriction (LDR) for a minimum period of 5 years.
   f. The documentation required for the emergency notification of all the local authorities for the Small Quantity generators (SQGs) or the Large Quantity Generators (LQGs).
   g. Having the right contingency plan for the Large Quantity Generators.

The existence of toxic waste material in a developing or a developing country is of no significant issue. The issue is for the effective handling of the waste materials so that the toxic does not actually effect any part of our environment, which eventually raises serious health concerns and threats for the life of human beings, plants and animals. The third world or the developing countries in this regard face serious challenges where the lack of efficient waste management plan, lack of technical facilities, insufficiency of training programs and training personnel and deficiency in the overall structure poses some serious issues. The need is to identify first the type of the waste and look for the separation of the toxic waste material from the non-toxic type, so that the non-toxic type could be further processed by putting it into the recycling phase for future reusability. There are plenty once the proper infrastructure is inducted into the waste management system of a developing country where people would enjoy such healthy life with no risk of environment or the pollution adding any type of threats to their health.

THE TWO CASES OF WASTE MANAGEMENT: URANIUM AND DISPOSAL OF WASTES FROM TANNING INDUSTRY

Case1: Prediction of Uranium Transport in an Aquifer at a Proposed Uum In situSitu Recovery Site: Geochemical Modelling as Decision-marking Tool.

Uranium deposits occurs as roll-fronts. In the United State these deposits are found in states such as Colorado and Utah. Uranium exists in two chemical states. Uranium hexavalent (the reduced state) is soluble and is found dissolved in ground water. Uranium tetravalent (the oxidized state) is insoluble and immobile existing as solid deposits. This oxidized Uranium is mined using a noninvasive pross –the Inin situ Recovery mining process (ISR).

This method involves drilling boreholes to the target mineral deposits and pumping solution meant to dissolve the Uranium ore. The ground water properties before and after the mining must remain same. The major c however, however, is the determination uranium levels in the ground water after mining. This paper focuses on the use of hydrous ferric oxides as a sorption for uranium to determine the movement of uranium.

Uranium is a radioactive element. It's also toxic making it dangerous to all living being. In humans, dissolved Uranium is attributed to nephritis, where Kidneys become inflamed. It also causes osteoporosis –wearing of bones – by reducing calcium, leading to fragile bones.

To form a uranium rich roll-front, surface water leaches uranium rich rocks and volcanic ash. The water containing dissolve uranium flows through permeable sand stones and aquifers. Finally, uranium is then deposited in a reducing environment within the migrating roll-front. The dissolved uranium is reduced to insoluble solid deposits. ISR mining involves a reverse process where the solids are re-dissolve in a solution of oxygen and carbon dioxide. In essence, oxygen and carbon dioxide oxidize the uranium tetravalent to the hexavalent which is readily soluble.

To obtain Uranium, the solution is pumped to a treatment plant. Recovery of the mineral is based on solvent extraction. An ion exchange medium collects the uranium ores which are stripped using hydrogen peroxide solution. A liquid ion-exchange mechanism is also used to recover the uranium. In the ISR site uranium is precipitated on pyrite or organic carbon, which was deposited under reducing conditions. Presence of HFO in the surrounding areas (down-gradient), creates an oxidizing environment. This leads to the sorption of uranium on the HFO minerals. The flow pattern of uranium is altered to form reduced to oxidized after mining. Meaning the ground water flows to the oxidized regions. In mining sites this movement detected by digging monitoring wells 125 meters from the main ISR shaft.

Mining is seen to adversely contaminate the groundwater from roll-fronts. This ground water may find its way to surrounding normal boreholes. Sufficient regulation in mining uranium is necessary. The minimal allowable concentration of radionuclides imposed by the United States Environmental Protection Agency is 30 micrograms per one liter of water. Most, ground water surrounding mining sites meets this s However, However, reactions at the roll-front lead to high concentration of radium, radon and uranium. This poses serious concerns to use of water for agriculture and domestic in the surrounding areas.

We will examine the consequences of Uranium tetravalent adsorption and absorption on Hydro ferrous oxides (HFOs). This are minerals characterized by being highly porous and are oxidizing zones. This zones could be responsible for Uranium movements. The movements are also influenced by the concentration of carbon dioxide, calcium and oxygen. To analyze the transport of uranium in ground water modelling is done using computer programs. The determination of the transport of uranium is necessary to influence decision making, enforce environmental rules and risk evaluation.

MATERIAL AND METHODS

The study took place at Dewey-Burdock ISR mining located at Edgemont,

Southern Dakota. Methods used are generally applicable to ISR mining boreholes elsewhere. During the initial mining steps, samples of the aquifer were collected. This served to analyze the composition of calcium and Hydro ferrous oxide using the Breit and Goldhaber methodology. Proven methods of analysis were also used to examine the surrounding uncontaminated groundwater.

The study of water bodies is based on geochemical modelling method used since the 1960s. This method enables an enhanced understanding of the interaction of rocks with water. Geochemical modelling, the principal method in hydrochemistry, is applicable in small laboratory studies. It's also expandable, for application in regulatory field studies. This method is used to determine the following:

- Transformation, and formation of complex inorganic molecules in a solution
- Saturation levels required for dissolution and formation of mineral precipitates.
- How cation species are exchanged on the surface of minerals through adsorption, and desorption.
- How single phase and multiphase reaction, and reactions involving organic materials take place.
- The changes in the chemical properties of a solution when redox conditions and temperature are varied.

Limited information exists on natur reactions,reactions; therefore, thermodynamic reactions form the premise behind geochemical modelling. Here, computers solve simultaneous equations of mass balance and mass balance. In this nonlinear system the assumption is that, an equilibrium already exists. Various computer programs such as the WATERQ4F, PHEREEQC and MINTEQA2, Geochemist's workbench are used to solve these complex thermodynamics.

The study employed the PHREEQC program to model specific reaction taking place in site's aquifer. When working at the laboratory level, this program is also suitable. It's widely applied to model, and simulate reaction, and natural process. The programing language that runs the program is C. The problem (query) forms the input of this program. It's a file with specific Keywords and data. Keywords are special variables such as; the solution spread, surface, transport and equilibrium phases.

More commands and mathematical functions are also possible. For better insight refer to the product manifest by Parkhurst and Appelo.

- Equilibrium phase: Refers to the different types of minerals, and gases reacting.
- Solution spread: Specifies the number of aqueous solutions used.
- Surface: Refers to the properties of the collection of all surfaces.
- Transport: Prescribes the standards for dispersive-reactive transport. It also includes parameters for porosity.
- End: This command is included to stop the programs.

By solving the mass action, and mass balance equations using the program the final chemical compositions of a solution under investigation is determined. The solved model becomes the Forward Geochemical model. The forward model is pooled with the groundwater flow model. This creates a Reactive Transport model. The essential feature of this model is the flow path. Through discretization the flow path is separated to discrete cells. To this individual smaller portions, the geochemical model is applied to determine composition of the final solution. The calculated results (specific compositions) are added to the next discrete cell. Again, the geochemical equation is solved by the computer to determine the final composition. This process is called the flow model. It's repeated to the cells in the flow path according to a prescribed number of steps – called "shifts"

In the study, the flow path consisted on the ISR zone divided into 8 cells followed by the HFO zone consisting of 32 cells. At a certain point in the flow path a monitoring point was created to monitor the groundwater. Flow rate for ground water was taken as 5m per year. One year was the approximate time required for ground water to penetrate one cell. In the flow path monitoring stations are drilled to sample the moving water from the mining zone. Contaminated water from the ISR mine can take up to twenty-five years to reach sampling points. Initial water from the ISR zones was found to be more abundant in oxygen, uranium and carbon dioxide, unlike the water from the monitoring wells. After mining uranium the clean groundwater is pumped to restore the same composition of this elements in the ISR zone. This is done before monitoring is carried out as specified by the EPA. In geochemical modelling the water used was of the same composition as the surrounding water. Various variables were used to predict the movement of uranium. This included variation of:

- The database
- Concentration of chemicals in ground water after mining activities
- Quantity of iron (HFO) surrounding the aquifer
- Level of calcium of rocks in the flow path (down-gradient)
- Concentration of carbon dioxide in the ISR zone after mining

Databases used to model reactions by the program included:

- The updated PHEEQC
- Wateq4f

The first database has recent data and is continuously updated with information on uranium complexes. Information on Magnesium-uranium and calcium-uranium carbonates is also available from the database. The second database came originally with the program.

To predict the sorption rates of uranium the water recovered after mining activities was analyze. Due to restoration the uranium levels and oxygen levels were similar to local gro. However, However, the composition used for the site was based on water from other sites. Fortunately, the rocks form the site were available for analysis.

The hydro ferrous oxide concentration was obtained by measuring the concentration of iron, and converting it using the method devise by Appelo and Postma. For the study, the uranium levels were sensitive to the calcium saturation levels. The simulations were conducted both at full saturation and at unsaturation. The concentration of CO2 was varied during the simulation run. High levels of carbon dioxide were adopted for the ISR zone. The timeline was based on the approximate time required to reach the monitoring well, which was twenty-five years.

After the simulation runs, the expected concentration of Uranium in the monitored water varied significantly due to the differences in the datab. However, However, the concentration of uranium was proportional to the concentration of iron in HFO. At zero sorption the concentration of uranium matched that of uncontaminated local groundwater.

The concentration of uranium in the moving water at the monitoring well after 25 years was proportional to the concentration of iron in the HFO. Decreasing the concentration of iron lead to reduced uranium. This change in concentration can be attributed to the process of absorption, adsorption and desorption (of uranium from the ISR zone) on the HFO mineral.

The update database, when used in simulation, lower levels of uranium at the monitoring wells was observed. The water database showed higher concentration of uranium. The plausible reason for this, is due to data on calcium carbonate complexes missing in the Wateq4f database. Calcium carbonate complexes increase the solubility of uranium thereby limiting sorption. Influence on other elements such as magnesium on the level of uranium was simulated. Th, however, however, remained unchanged due to the minimal influence on sorption.

For the updated database the effect of the variation of the calcium concentration was easily noticeable. This was attributed to the reliance of sorption to calcium uranium carbonates. As the concentrations decrease more uranium was adsorbed or absorbed to the HFO decreasing its rate of movement in the flow path. This resulted to a lower level of uranium in the final ground water solution.

Carbon dioxide

As noted earlier, the solution used in the extraction of uranium contains Co2 and O2. Thus a higher concentration in the final solution is plausible. To study the effect of carbon dioxide on sorption was studied by varying its concentration whilst maintain the concentration of calcium and HFO. In the flow path, the concentration of uranium first dropped in the first cell then increased in subsequent cells before, drastically reducing. In the first cells after the ISR zone the concentration of uranium was at a minimum. This likely resulted due to high concentration of CO2 which created an acidic solution. This reacted with all the calcium leading lower sorption rates. As more CO2 dissolved the pH of the water started increasing and more sorption took place. Also, the absorbed uranium in the first cells was released back to the flow path, gradually increasing uranium concentrations away from the ISR zone. The drastic reduction is plausible due unexamined geochemical processes.

- It's possible to predict of the concentration of uranium at stages in the flow path based on simulations. This is helps to determine if, the requirements after restoration are obtainable.
- T, however, however, to a larger extent relies on theoretical estimates of HFO and presumed sorption capacities.
- By evaluating the effects of sorption by the HFO, uranium mining sites can be analyzed effectively and assist in determining final conditions.

To increase the credibility of forecasts:

- Analyze real water from the ISR site after the clean-up process.
- Employ real gradient mineralogy concentration of iron and calcite
- Carry out batch or column studies of the actual sorption potential of the flow path.

https://www.hindawi.com/journals/er/2011/805187/

OVERVIEW OF TANNING AND WASTE DISPOSAL TECHNIQUES

Leather tanning is without doubt one of the oldest human activities. In the beginning, skins obtained from hunting and livestock breeding could be used for clothing or tents, but in conditions of low temperatures they became stiff. When the temperatures went up they would rot. It was probably then that attempts were made to render them more flexible and stronger by rubbing in animal fats. This was the first tanning process.

Leather industry is one of the fast-growing traditional sectors with high environmental challenges. Leather manufacture deals with the conversion of putrescible material such as skin/hide to non-putrescible material. During the conversion, series of chemical processes and physical operations are carried out to attain the final desired properties of leathers.

Pre-tanning, tanning, and post-tanning are the major steps involved in the leather processing. One of the major challenges of leather processing is generation of high solid and liquid wastes that makes it more an environmental concern.

Pre-tanning is the process of treating of skins and hides of animals to produce leather. A tannery is the place where the skins are processed. Tanning hide into leather involves a process which permanently alters the protein structure of the skin, making it more durable and less susceptible to decomposition and also possibly coloring it. Here, the raw materials go through thorough cleaning and removal of unwanted materials. Steps involved include, soaking, liming, reliming, fleshing, and unhairing, pickling are involved in pre-tanning. Solid wastes such as hair, lime sludge, interfibrillary materials, and adipose tissues are the major components during this step. Liquid wastes are generated in all unit processes contributing to the chemicals used in the respective steps. Due to heavy pollution load, greener leather processes have been adopted.

The steps involved are:

- Preservation- the hide/skin is treated with a method which renders it temporarily imputrescible.
- Soaking - water for purposes of washing or rehydration is reintroduced.
- Liming - unwanted proteins and "opening up" is achieved.
- Unhairing - the majority of hair is removed.
- Fleshing- subcutaneous material is removed.
- Splitting - the hide/skin is cut into two or more horizontal layers.
- Reliming - the hide/skin is further treated to achieve more "opening up" or more protein removal.
- Deliming - liming and unhairing chemicals are removed from the pelt.
- Bating - proteolytic proteins are introduced to the skin to remove further proteins and to assist with softening of the pelt
- Degreasing - natural fats/oils are stripped or as much as is possible from the hide/skin
- Frizing- physical removal of the fat layer inside the skin. Also

similar to Slicking.

- Bleaching - chemical modification of dark pigments to yield a lighter coloured pelt.
- Pickling- lowering of the pH value to the acidic region. Must be done in the presence of salts. Pickling is normally done to help with the penetration of certain tanning agents, e.g., chromium (and other metals), aldehydic and some polymeric tanning agents.

The acidity of hides once they have finished pickling will typically be between pH of 2.8-3.2. At this point the hides are loaded in a drum and immersed in a float containing the tanning liquor. The hides are allowed to soak (while the drum slowly rotates about its axle) and the tanning liquor slowly penetrates through the full substance of the hide. Regular checks will be made to see the penetration by cutting the cross section of a hide and observing the degree of penetration. Once an even degree of penetration is observed, the pH of the float is slowly raised in a process called basification. This basification process fixes the tanning material to the leather, and the more tanning material fixed, the higher the hydrothermal stability and increased shrinkage temperature resistance of the leather.

Crusting

Crusting is when the hide/skin is thinned, retanned and lubricated. Normally, this is often a coloring operation is included in the crusting sub-process. The chemicals added during crusting have to be fixed in place. The culmination of the crusting sub-process is the drying and softening operations. Crusting may include the following operations

- Wetting back - semi-processed leather is rehydrated.
- Summing-water is squeezed out the leather.
- Splitting - the leather is split into one or more horizonta layers.
- Sshaving - the leather is thinned using a machine which cuther fesfibtralisationNeutralization- the pH of the leather is adjusted to a value between 4.5 and 6.5.
- Retanning - additional tanning agents are added to impart properties.

- Dyeing - the is colouredcot-liqouringliquoring- fats/oils and waxes are fixed to tther fibresfibers.
- Filling - heavy/dense chemicals that make the leather harder and heavier are added.
- Stuffing - fats/oils and waxes are added b the fibresfibers.
- Stripping- superficially fixed tannins are removed.
- White the colourcolor of the leather is lightened.
- Fixation - all unbound chemicals are chemically bonded/ trapped or removed from the leather
- Setting - area, grain flatness are imparted and excess water removed.
- Drying - the leather is dried to various moisture levels (commonly 14-25%).
- Conditioning - water is added to the leather to a level of 18-28%.
- Softening - physical softening of the leather by separating tther fibresfibers.
- Buffing - abrasion of the surfaces of the leather to reduce nap or grain defects.

Solvent-based post-tanning is carried out to reduce the pollution and maximize the chemicals exhaustion. Though several alternate cleaner leather processes exist, no holistic approach has been made till date. Furthermore, conventional leather processing methods have found an impeccable place among the tanners due its mis-unmatching physical and chemical properties to the leathers. Therefore, it would be highly appreciable if wastewater treatment becomes a relevant measure to combat the pollution load. Primary, secondary, and tertiary treatments are practiced to treat the wastes. Biological treatment methods are the fast-growing technology due to their eco-friendly and recycle method.

Researchers also work to develop natural-derived materials from plants and microorganisms for tanning such as dialdehyde cellulose, dialdehyde starch, dialdehyde alginic acid,dialdehyde pectin, and scleraldehyde from fungal species (Schizophyllum commune). Post-tanning comprises retaining, dyeing, and fat liquoring to impart special properties to leathers as a functional requirement. The waste stream is composed of mixed wastes that make it more difficult during wastewater treatment. The composition of tannery wastewater is shown below.

Pollutants	**value**
Total suspended solids	50mg/mL
Total suspended solids Sulfide	1.0mg/mL
Chromium(VI)	0.1mg/ml
Total chromium	0.5mg/ml
Chloride	1000mg/mL
Sulfate	300mg/mL
Ammonia	10mg/mL
Oil and grease	10mg/mL
Total nitrogen	10mg/mL
Total phosphorous	2mg/mL
Phenols	0.5mg/mL
Total coliform bacteria	400MPN/100ml
Temperature increase	<3°C

Source: https://www.intechopen.com/books/management-of-hazardous-wastes/

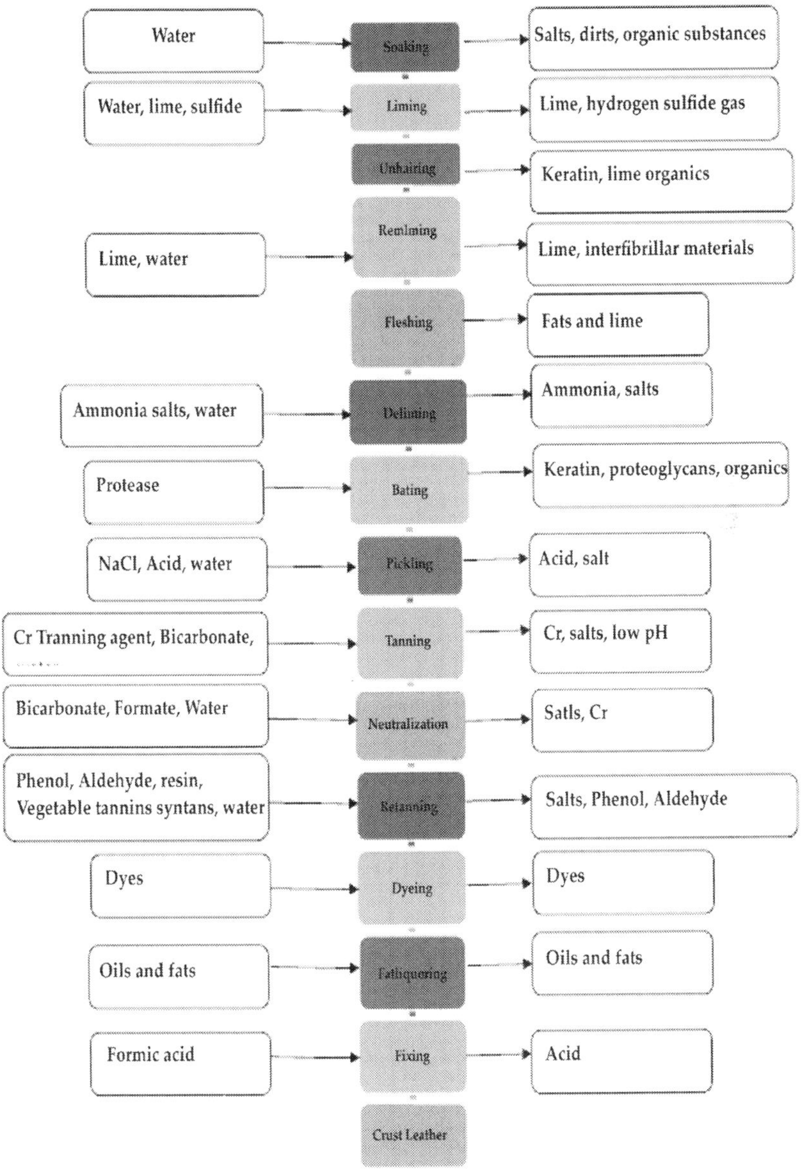

Source: https://www.intechopen.com/books/management-of-hazardous-wastes/

VEGETABLE TANNING

Vegetable tanning is the most traditional way of tanning hides. Using Italian artisans, most of the process is done by hand. Unlike chrome tanning, vegetable tanning can take up to 40 days to produce a piece of dyed leather. Vegetable tanning uses natural ingredients such as the bark of chestnut trees to tan the leather. This means the finished product color is usually warm, rich tones of natural browns. Vegetable tanning is usually done by tanneries that have a rich heritage of tanning hides.

Advantages and Disadvantages of Vegetable tanning

- Only natural ingredients are used when drying the hides.
- Vegetable tanning is environmentally friendly; meaning any leather products that have been vegetable tanned can be recycled.
- Vegetable tanning is an age old tradition, so most tanneries have very skilled craftsmen producing and dying the leather.
- Due to the natural tannins used, vegetable tanned products actually improve with age and develop a patina.
- Each leather product that is dyed using vegetable tanning is completely unique.
- The colors that vegetable tanning produces are rich and warm tones that look completely natural.

The disadvantages

The process of vegetable tanning takes a long time, and can take up to 60 days.

- Vegetable tanning doesn't react well with water and can stain easily.
- Products that have been vegetable tanned are more expensive.
- The colors you can produce from vegetable tanning are limited.
- Direct heat can cause vegetable tanned products to shrink or crack

Advantages and Disadvantages of Chrome tanning

- Quick and easy to produce, usually only taking up to a day.

- Stain resistant and water can roll off the surface easily.
- Soft and supple to the touch.
- Color remains the same throughout the products life.
- It's cheaper to buy than vegetable tanned leather, which means it is also easier to find.
- It has a high degree of thermal resistance.

***Disadvantages*:**

- Chrome tanning is very bad for the environment.
- It's produced with little skill and very often mass produced.
- It doesn't wear very well or last long and can crack after a few months of use.
- Chrome tanning often smells of chemicals.
- It doesn't appear (neither is it) very natural.

The diagram below summarizes the different methods used in tannery wastewater treatment.

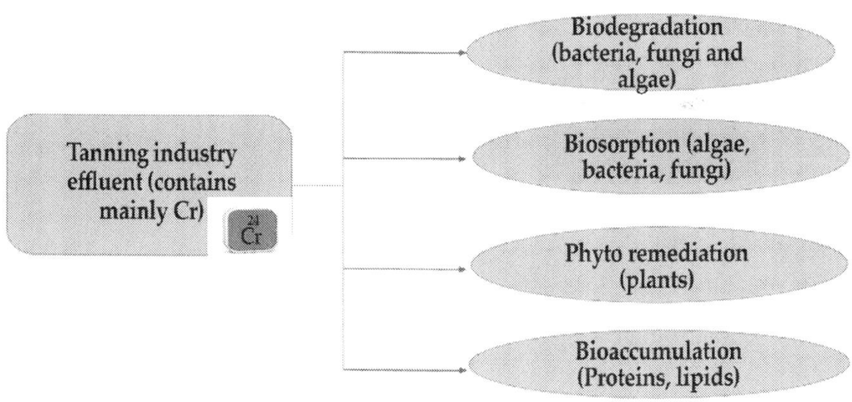

Source: https://www.intechopen.com/books/management-of-hazardous-wastes/

APPLICATION OF MICROORGANISMS IN TANNERY WASTEWATER TREATMENT

The role of microorganisms is gaining importance in different applications due to their versatile abilities. Availability of single cell microorganism is largely found in the ecology system. These organisms are mainly found in the soil systems, water, and intestinal tract of living system. Microorganisms are classified based on the sources of energy and carbon. The major classifications are autotrophs and heterotrophs. Autotrophs are further classified into photoautotrophic and chemoautotrophic based on energy source as light and inorganic oxidation-reduction reaction, respectively. Heterotrophs are further classified into chemoheterotrophic and photoheterotrophic based on their energy source, organic oxidation-reduction reaction, and light, respectively. One of the unique aspects of their living organisms is their size, regular supplements of nutrition, water, and optimum physiological condition to grow and multiply the cells.

During secondary treatment of leather wastes, microorganisms commonly used result in the generation of biomass sludge.

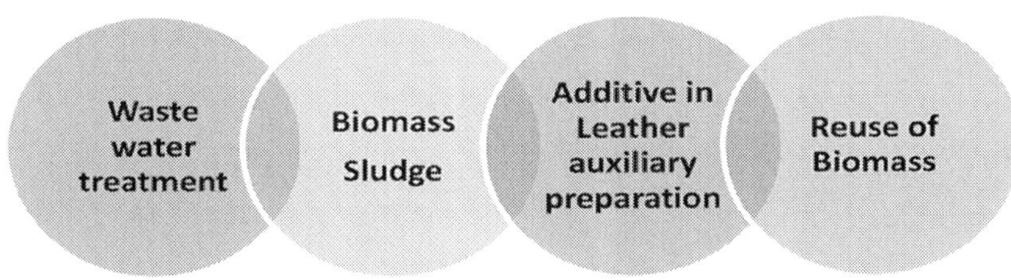

Source: https://www.intechopen.com/books/management-of-hazardous-wastes/

Bacteria

Bacteria are single microorganisms and are in small size with different shapes. For bacteria to grow, they require nutritional supplements like any other living organisms. Nutritional requirements are obtained by decomposing organic material and aids in multiplying the cell. Hence, bacteria are efficiently used for decaying organic wastes present in the leather waste stream. Leather wastes during pre-tanning processes are chiefly made up of organic constituents such as collagen, elastin, reticulin, and other nutritional supplements. Aerobic and anaerobic systems are practiced in treating wastewater in the secondary treatment plant. Chemolithotrophic bacteria are used to absorb the toxic gas.

During this process, bacterial cells are purged to undergo biological oxidation. Bacteria tend to degrade the organic component in the waste stream and are used as nutritional supplement. This respiration process utilizes oxygen and exits the carbon dioxide. After the process the wastewater treatment plant will end up with a large amount of activated sludge (bacterial cell mass). Very often the activated sludge is condensed and treated in large-scale anaerobic digesters designed for further treatment via anaerobic bacteria. However, there should be very careful steps taken toward the disposal of bacterial sludge. Since, it is mainly composed of proteins and carbohydrates which were used as a raw material for preparing leather chemicals. Protein fillers and protein synthetic tanning agents are possible areas to venture for the recycling bacterial sludge.

Below is a list bacteria used in tannery wastewater treatment.

Bacteria	**process**
Bacillus subtilis	Biosorption of Cr
Bacillus sp	Livestock
Arthrobacter sp.	ultrafiltration/microfiltration lead
Escherichia coli	Removal of chlorphenol

Algae

These are significant organisms for biological purification of wastewater. They are important because they can be able to accumulate plant nutrients, heavy metals, pesticides, and organic and inorganic toxic substance. The application of algae in biological wastewater treatment has gained a lot of importance in the recent years. Construction and energy costs are highly lower and the land requirement is not up to that of facultative pond in constructed wet lands. Algae are commonly used to absorb sectional waste from the composite waste liquors. They are chiefly used for the absorption of heavy metals due to their high tolerance and absorption. Moreover, the large surface area/volume ratio makes it as one of the important biological materials for wastewater treatment.

Algae are composed of polysaccharides that make convenient for the absorption of heavy metals. Their high tolerance to heavy metals and growth behavior of autotrophs and heterotrophs make them one of the best choices. Biosorption of cadmium, chromium, and copper ions by other blue green algae and Chlorella sp. is known well. Application of algae in absorption of chromium leads to the generation of sectional solid waste. It would be mere conversion of liquid into solid waste; hence, further intensive research needs to be carried out in order to effectively and efficiently reuse the solid waste. Since, solid waste is composed of carbohydrates and metals; it can be used as a raw material for preparing chrome-based retaining agents. It can be further investigated as one of the chemical components during post-tanning processes with chemical modifications.

Below is a list of algae used in tannery wastewater treatment.

Algae	**Process**
Phormidium bohneri	Removal of nitrogen and phosphorus
Spirulina (Arthrospira)	Removal of nitrogen and phosphorus
Scenedesmus sp.	Immobilized to remove nutrients (P and N)
Spirogyra condensate	Biosorption of chromium
Scenedesmus sp.	Bioflocculation

Fungi

Fungi are multicellular organisms unlike bacteria which are unicellular organisms and are abundantly present in wastewater stream. These fungi are found to be effectively used in the removal of nutrient source from the wastewater Fungi have a critical role in treatment at very high acidic pH. Though acid-resistance bacteria exist, there is still a challenge for the bacterial treatment at low pH. Hence, fungi can be used in the acidic medium to break down the organic constituents. Moreover, fungi have the ability to trap and absorb the material and aid in the metabolism. Enzymes are also secreted by some fungi that help in the biodegradation of organic compounds present in the waste.

- Application of plants in bioremediation

Phytoremediation

This is a branch of bioremediation that deals with the utilization of plants for the wastewater treatment. Halophytes and aquatic plants are more commonly used for the tannery waste treatment. The major selections of plants are based on the availability and growth surrounding the tannery plants. This is based on the hypothesis that plants are resistant to the environmental parameters and resistant to the pollutants discharged from the tanning industry.

Constructed wetland

This is an artificial wetland widely used for treatment of municipal and industrial wastewater. Constructed wetland system is a hybrid technology including phytoremediation, soil filtration, and microbial processes and has been reported to enhance overall biological performance by complex interactions inside the ecosystem. In the past decade, researchers have attempted to study the potential of constructed wetland for the treatment of tannery effluent.

Cristina et al., a research group, have investigated four different plants Canna indica, Phragmites australis Iris pseudacorus. Typha latifolia, and Stenotaphrum secundatum. These plants are vegetated in wetlands for the treatment of tannery wastewater. The plants are chosen based on the growth and availability surrounding the tannery effluent contaminated lands. Among the chosen five plants P. australis and T. latifolia are effectively involved in the waste treatment. It might be understood that there would be growth of plants surrounding tannery effluent–contaminated plants. However, it is not mandate that all plants have the ability to effectively participate for the effluent treatment.

Daniel et al. have carried out on water hyacinth (Eichhornia crassipes) for the removal of chromium from tannery effluent in constructed pond system. Water hyacinth is free-floating aquatic plants that originated from Amazon, South America. The samplings are collected in the Awash River, Ethiopia, for the treatment. Many other plant species are also involved in the phytoremediation; however, the removal of heavy metals and nutrients should be considered for further applications.

Wastewaters, including domestic and industrial wastewaters, contain a variety of compounds. Some of the common compounds present in wastewaters and in other effluents are phenolic compounds, hydrocarbons, dyes, endocrine disrupting compounds, and pesticides.

Landfills generate large amounts of leachates containing high levels of organics and ammonia nitrogen. These substances with others like phenols and hydrocarbons can be a major source of contamination of the groundwater. Indeed, the variety of contaminants in landfill leachates, their synergistic and antagonistic effects as well as their physicochemical properties make them serious toxicants, which may survive different treatments.. Landfill leachates exhibit consequently high toxicity levels. The efficiency of fungal remediation of landfill leachates has been proved on Lentinus tigrinus, ,Trametes trogii, and P. chrysosporium. The strains were able, via their extracellular enzymes, to reduce organics (chemical oxygen demand (COD), phenols, and hydrocarbons) as well as toxicity, for twofold diluted LFL. However, raw LFL caused growth inhibition and enzyme secretion reduction, indicating the sensitivity of these strains to high levels of toxic compounds such as phthalates and phenol derivatives.

Tigini et al. reported that autochthonous and allochthonous fungal strains were efficient in LFL treatment, showing a complete spectrum of action and being able to significantly reduce the wastewater toxicity for all the tested strains. Thus, Porostereum spadiceum showed the best activity with 40 % of decolorization within 1 week.

The contamination of soil, water, ground water, and air with toxic chemicals is one of the major environmental problems, faced by the world today. With the intensive industrialization and the extensive use of pesticides in agriculture and chemicals in various fields, the environmental pollution with organic compounds has become a serious threat. Indeed, pollution of aquatic system and soil is a worldwide problem that can result in accumulation of toxic chemicals in food chains and also harm the flora and fauna. Hence, environmental pollution with hazardous wastes containing recalcitrant chemicals, being often xenobiotic compounds, has become one of the major ecological problems, with an increasing awareness around the world.

Xenobiotics are chemicals that are "foreign to the biosphere" and may become available to microorganisms in different environmental compartments, depending on their fate in air, water, soil, and sediments. The most hazardous compounds are persistent in the environment and are carcinogenic and/or mutanogenic. Household chemicals, pharmaceuticals, and other consumables as well as biogenic hormones are released into the environment after passing through wastewater treatment processes, which are not designed for their removal.

The quick rise of industrial activities has extremely increased the release of toxic effluents into water bodies along with ground water. The pollution resulting in the release of these compounds causes disturbance to the natural bodies and their ecosystems, leading to climatic changes, water level reduction, and other negative impacts. On the other hand, there are increasing concerns about potential adverse health and ecological effects resulting from the production, the use, and the disposal of numerous chemicals that otherwise offer improvements in human life and economic activities. Thus, a huge amount of hazardous organopollutants is produced annually over the world and only 10% of these are disposed safely.

The main sources of xenobiotics are wastewater and solid wastes released from the industries directly, such as phenols, plastics, hydrocarbons, paints, dyes from textile mills, pesticides and insecticides from agricultural industries and paper and pulp mills, or indirectly, including pharmaceuticals, especially the group of endocrine disrupters, and pesticide residues. These chemicals include biopolymers synthetic polymers, polycyclic aromatic hydrocarbon, pentachlorophenols (PCP)).

Unlike the naturally occurring organic compounds that are readily degraded upon introduction into the environment, some of these synthetic chemicals are extremely resistant to degradation by native microorganisms.

Degradation of such compounds by physical and/or chemical processes is costly and often produces undesirable products which are toxic. Biological methods, being eco-friendly and cost cheap techniques, were proposed for xenobiotic degradation purposes in order to overcome these problems. Compared to bacteria, most of the fungi are robust organisms and generally more tolerant to high concentrations of pollutants. It explains why they have been extensively investigated since the mid-1980s for their bioremediation capacities. White-rot fungi (WRF) constitute an eco-physiological group comprising mostly of basidiomycetes and litter-decomposing fungi.

https://www.intechopen.com/books/management-of-hazardous-wastes/ white-rot-fungi-and-their-enzymes-as-a-biotechnological-tool-for-xenobiotic-bioremediation

Solid waste residues can be domestic wastes, including food, paper, and garden wastes; waste from council activities associated with servicing residential areas: street sweepings, tree lopping, parks and gardens, and litter bins; and waste from institutional, commercial, industrial activities, generally containing higher proportions of metals and plastics than domestic wastes. They also can be derived from demolition and building activities, which contain high proportions of inert material (concrete, bricks) and low proportions of other materials. Many xenobiotic compounds are released from municipal solid waste and may be found in the leachates and the gaseous phase of landfills.

The quick rise of industrial activities has extremely increased the release of toxic effluents into water bodies along with ground water. The pollution resulting in the release of these compounds causes disturbance to the natural bodies and their ecosystems, leading to climatic changes, water level reduction, and other negative impacts. On the other hand, there are increasing concerns about potential adverse health and ecological effects resulting from the production, the use, and the disposal of numerous chemicals that otherwise offer improvements in human life and economic activities. Thus, a huge amount of hazardous organopollutants is produced annually over the world and only 10% of these are disposed safely.

The main sources of xenobiotics are wastewater and solid wastes released from the industries directly, such as phenols, plastics, hydrocarbons, paints, dyes from textile mills, pesticides and insecticides from agricultural industries and paper and pulp mills, or indirectly, including pharmaceuticals, especially the group of endocrine disrupters, and pesticide residues. Unlike the naturally occurring organic compounds that are readily degraded upon introduction into the environment, some of these synthetic chemicals are extremely resistant to degradation by native microorganisms.

Degradation of such compounds by physical and/or chemical processes is costly and often produces undesirable products which are toxic. Biological methods, being eco-friendly and cost cheap techniques, were proposed for xenobiotic degradation purposes in order to overcome these problems. Compared to bacteria, most of the fungi are robust organisms and generally more tolerant to high concentrations of pollutants. It explains why they have been extensively investigated since the mid-1980s for their bioremediation capacities. White-rot fungi (WRF) constitute an eco-physiological group comprising mostly of basidiomycetes and litter-decomposing fungi.

Recently, there has been a great interest in white-rot fungi and their ligninolytic enzymes, including laccase, manganese peroxidase (MnP) and lignin peroxidase (LiP), for the degradation of a wide range of xenobiotics. These xenobiotics include phenols, hydrocarbons, dyes and textile effluents, pharmaceuticals including the endocrine-disrupter compound, pulp and paper mills pesticides, and insecticides. Some studies reported the effect of ligninolytic enzymes on degradation of other organopollutants like PCB, chlorinated phenols including trichlorophenols, and xenobiotics in landfill leachates, solid wastes, and munitions.

- White-rot fungi

Because of the non-specific nature of fungal lignin depolymerization, white-rot fungi can degrade a wide range of persistent environment pollutants even the insoluble chemicals. These microorganisms generally act via the extracellular ligninolytic system showing good potential applications in chemical, agro-food, paper, textile, and cosmetic industries. This group may be a useful and a powerful tool for bioremediation purposes thanks to fungal capacities to degrade many xenobiotic substances.

Source: https://www.intechopen.com/books/management-of-hazardous-wastes/white-rot-fungi-and-their-enzymes-as-a-biotechnological-tool-for-xenobiotic-bioremediation

The expression of these enzymes depends on the strain itself: some white-rot fungi produce LiP and MnP, but not laccase, while others produce MnP and laccase, but not LiP, acting simultaneously or separately on xenobiotics released from the environment. The potential of white-rot fungi can be harnessed thanks to emerging knowledge of the physiology and the morphology of these organisms. This knowledge could be transformed into reliable and robust waste treatment processes. The importance of high extracellular levels of these enzymes to enable the efficient degradation of recalcitrant compounds under in vivo conditions relates to the sorption and complexation of enzymes in soil and the probable loss of their activity once externalized.

The lignin degradation system consists on peroxidases, H2O2-producing enzymes, veratryl alcohol (3,4-dimethoxybenzyl alcohol), oxalate, and manganese. LiP and MnP are glycosylated heme proteins that couple the reduction of H2O2 to water with the oxidation of a variety of substrates. It was reported that Phanerochaete chrysosporium, producing simultaneously both MnP and LiP, was able to degrade many xenobiotics and recalcitrant compounds.

ENZYMATIC SYSTEM INVOLVED IN XENOBIOTIC BIODEGRADATION

Enzymes are biological catalysts that facilitate the conversion of substrates into products by providing favorable conditions that lower the activation energy of the reaction. An enzyme may be a protein or a glycoprotein and consists of at least one polypeptide moiety. The regions of the enzyme that are directly involved in the catalytic process are called the active sites. An enzyme may have one or more groups that are essential for catalytic activity associated with the active sites through either covalent or noncovalent bonds; the protein or glycoprotein moiety in such an enzyme is called the apoenzyme, while the nonprotein moiety is called the prosthetic group. The combination of the apoenzyme with the prosthetic group yields the holoenzym.

Enzyme classification

The identification of a particular enzyme is possible through its enzyme commission (E.C.) number. The assignment of E.C. numbers is described in guidelines set out by the International Union of Biochemistry. All known enzymes fall into one of these six categories. The six main divisions are the oxidoreductases, the transferases, the hydrolases, the lyases, the isomerases, and the ligases (synthetases). Oxidoreductases catalyze the transfer electrons and protons from a donor to an acceptor. Transferases catalyze the transfer of a functional group from a donor to an acceptor. Hydrolases facilitate the cleavage of C–C, C–O, C–N, and other bonds by water. Lyases catalyze the cleavage of these same bonds by elimination, leaving double bonds (or, in the reverse mode, catalyze the addition of groups across double bonds). Isomerases facilitate geometric or structural rearrangements or isomerizations. Finally, ligases catalyze the joining of two molecules.

Microbial enzymes in Bioremediation

Microbial Oxidoreductases

The detoxification of toxic organic compounds by various bacteria and fungi and higher through oxidative coupling is mediated with oxidoreductases. Microbes extract energy via energy-yielding biochemical reactions mediated by these enzymes to cleave chemical bonds and to assist the transfer of electrons from a reduced organic substrate (donor) to another chemical compound (acceptor). During such oxidation-reduction reactions, the contaminants are finally oxidized to harmless compounds (ITRC 2002).

The oxidoreductases participate in the humification of various phenolic substances that are produced from the decomposition of lignin in a soil environment. In the same way, oxidoreductases can also detoxify toxic xenobiotics, such as phenolic or anilinic compounds, through polymerization, copolymerization with other substrates, or binding to humic substances Microbial enzymes have been exploited in the decolorization and degradation of azo.

Many bacteria reduce the radioactive metals from an oxidized soluble form to a reduced insoluble form. During the process of energy production, bacterium takes up electrons from organic compounds and use radioactive metal as the final electron acceptor. Some of bacterial species reduce the radioactive metals indirectly with the help of an intermediate electron donor. Finally precipitant can be seen as the result of redox reactions within the metal-reducing bacteria.

Chlorinated phenolic compounds are among the most abundant recalcitrant wastes found in the effluents generated by the paper and pulp industry. These compounds are produced upon the partial degradation of lignin during pulp bleaching process. Many fungal species are considered to be suitable for the removal of chlorinated phenolic compounds from the contaminated environments. The activity of fungi is mainly due to the action of extracellular oxidoreductase enzymes, like laccase, manganese peroxidase, and lignin peroxidase, which are released from fungal mycelium into their nearby environment. Being filamentous, fungi can reach the soil pollutants more effectively than bacteria.

Water polluted with phenolic compounds can be de-contaminated by plants with the help of enzymes exuded by their roots. The plant families of Fabaceae, and Solanaceae are found to release oxidoreductases which take part in the oxidative degradation of certain soil constituents. Phytoremediation of organic contaminants has been generally focused on three classes of compounds: chlorinated solvents, explosives, and petroleum hydrocarbons.

Microbial Oxygenases

Oxygenases belong to the oxidoreductase group of enzymes. They participate in oxidation of reduced substrates by transferring oxygen from molecular oxygen (O2) utilizing FAD/NADH/NADPH as a cosubstrate. Oxygenases are grouped into two categories; the monooxygenases and dioxygenases on the basis of number of oxygen atoms used for oxygenation. They play a key role in the metabolism of organic compounds by increasing their reactivity or water solubility or bringing about cleavage of the aromatic ring. Oxygenases have a broad substrate range and are active against a wide range of compounds, including the chlorinated aliphatics. Generally the introduction of O2 atoms into the organic molecule by oxygenase results in cleavage of the aromatic rings. Historically, the most studied enzymes in bioremediation are bacterial mono- or dioxygenases. A detailed study of the role of oxygenases in biodegradation process is available.

Halogenated organic compounds comprise the largest groups of environmental pollutants as a result of their widespread use as herbicides, insecticides, fungicides, hydraulic and heat transfer fluids, plasticizers, and intermediates for chemical synthesis. The degradation of these pollutants is achieved by specific oxygenases. Oxygenases also mediate dehalogenation reactions of halogenated methanes, ethanes, and ethylenes in association with multifunctional enzymes.

https://www.ncbi.nlm.nih.gov/pmc/articles/PMC3168789/

Monooxygenases

These incorporate one atom of the oxygen molecule into the substrate. They are classified into two subclasses based on the presence cofactor: flavin-dependent monooxygenases and P450monooxygenases. Flavin-dependent monooxygenases contain flavin as prosthetic group and require NADP or NADPH as coenzyme. P450 monooxygenases are heme-containing oxygenases that exist in both eukaryotic and prokaryrotic organisms. The monooxygenases comprise a versatile superfamily of enzymes that catalyzes oxidative reactions of substrates ranging from alkanes to complex endogenous molecules such as steroids and fatty acids. Monooxygenases act as biocatalysts in bioremediation process and synthetic chemistry due to their highly region-selectivity and stereoselectivity on wide range of substrates. Majority of mono-oxygenase studied previously are having cofactor, but there are certain monooxygenases which function independent of a cofactor. These enzymes require only molecular oxygen for their activities and utilize the substrate as reducing agent.

Microbial dioxygenases

Dioxygenases are multicomponent enzyme systems that introduce molecular oxygen into their substrate. Aromatic hydrocarbon dioxygenases, belong to a large family of Rieske nonheme iron oxygenases. These dioxygenases catalyze specifically the oxygenation of wide range of substrates. Dioxygenases primarily oxidize aromatic compounds and, therefore, have applications in environmental remediation. All members of this family have one or two electron transport proteins preceding their oxygenase components. The crystal structure of naphthalene dioxygenase has confirmed the presence of a Rieske (2Fe–2S) cluster and mononuclear iron in each alpha subunit.

The catechol dioxygenases serve as part of nature's strategy for degrading aromatic molecules in the Environment. They are found in the soil bacteria and involved in the transformation of aromatic precursors into aliphatic products. The intradiol cleaving enzymes utilize Fe(III), while the extradiol cleaving enzymes utilize Fe(II) and Mn(II) in a few cases.

Microbial peroxidases

The lignin degradation system consists on peroxidases, H_2O_2-producing enzymes, veratryl alcohol, oxalate, and manganese. All of these enzymes are glycosylated heme proteins that couple the reduction of hydrogen peroxide to water with the oxidation of a variety of substrates. The redox potentials of LiP and MnP are higher than for others peroxidases; that is why they have been shown to oxidize chemicals that are not easy to be oxidized by other microorganisms. These chemicals include Polycyclic aromatic hydrocarbons (PAH), phenol and its derivatives, cyanide, TNT, and others. This finding was reported for the fungus P. chrysosporium, which has been shown to degrade many xenobiotics and recalcitrant compounds, both in soil and in liquid cultures, suggesting the attractive use of such fungus in bioremediation.

Lignin peroxidases (LiPs) belong to the family of oxidoreductases and were firstly described in the basidiomycete P. chrysosporium in 1983. This enzyme has been recorded for several species of white-rot basidiomycetes. LiP is dependent of H_2O_2, with an unusually high redox potential and low optimum pH. This enzyme is able to oxidize a variety of substrates including polymeric ones and has consequently a great potential for application in various industrial treatment processes.

Manganese peroxidases (MnPs) belong to the family of oxidoreductases. Following the discovery of LiP in P. chrysosporium, MnP secreted from the same fungus was found as another lignin-degrading enzyme and was secreted by almost all white-rot fungi. MnP catalyzes the oxidation of phenolic structures to phenoxyl radicals. The product Mn^{3+}, being highly reactive, complex with chelating organic acids, such as oxalate, lactate, or malonate. On the other hand, it was reported that MnP may oxidize Mn(II) without H_2O_2 and with decomposition of acids, and concomitant production of peroxyl radicals.

Microbial Laccases

Laccases which are blue multicopper oxidases, catalyze the monoelectronic oxidation of a large spectrum of substrates, for example, ortho- and para-diphenols, polyphenols, aminophenols, and aromatic or aliphatic amines, coupled with a full, four electron reduction of O_2 to H_2O. Laccases act on both phenolic and nonphenolic lignin-related compounds as well as highly recalcitrant environmental pollutants, and they can be effectively used in paper and pulp industries, textile industries, xenobiotic degradation, and bioremediation and can act as biosensors.

Some studies reported the identification of genes that are differentially regulated during fungal growth in the presence of different environmental polHowever, However, abiotic stress caused by many factors including water potential, temperature, and pH can influence the metabolism of the degradation process. Hence, considering bioremediation in soil, the conditions that favor fungal activity in soil, such as temperature, moisture, nutrient status, pH, and aeration, need to be optimized to promote metabolic degradation of xenobiotics. Magan et al. studied the effect of abiotic factors on the fungal degradation of pesticides by T. versicolor and P. chrysosporium for soil bioremediation purposes. In fact, the potential property of laccase is its highly non-specific nature of substrates.

Furthermore, the common presence of one or more substructures in the lignin molecule and in xenobiotics explains the ability of white-rot fungi to degrade such a wide range of environmental organic pollutants, even at high levels. Otherwise, it has been shown that laccase metabolizes these compounds without any net energy gain. Indeed, the oxidation of lignin is performed to access to wood polysaccharides, being their main energy source. This implies that the presence of lignin-cellulosic substrates is required to ensure the degradation of xenobiotic compounds.

Reuse and recycle of treated waste in tanning industry by-product utilization

The annual tannery generation is estimated to be more than 600,000 tones of waste. This poses a major challenge to the environment. It is reported that producing 150 kg of leather would require 1000 kg of raw hide and throw out 850 kg of solid waste to the environment. With the high discharge amounts in leather processing, there is a critical need to treat and recycle the tanning industry by-product.

HIERARCHY OF TANNING WASTE TREATMENT/MANAGEMENT

The order of tanning waste management starts from the top of the hierarchy by friendlier option and finish at the bottom of the hierarchy in accordance with using friendly alternative, preventing, reusing, and recycling, treating and dumping waste.

1. Using environmentally friendly alternatives: Using biodegradable fat-liquoring alternatives, innovative compounds, which are based on natural materials such as palm oil, are the most preferable option, making them easier to treat wastewater from the tanning sector. Furthermore, the use of innovative technology for leather-tanning process with the goal of reducing hazardous waste should be chosen.
2. Preventing: Eliminating fat-liquoring agents such as chromium III, arsenic, cadmium, and lead is also recommended in tanning industry in order to prevent hazardous contaminants.
3. Reusing: The encouragement of reusing waste and residue for the reproduction or making new products is considered as the third option.
4. Recycling: Applying green technologies to treat waste by applying some energy conversion technologies such as biomethanation, bioenergy production and other feasible technologies.
5. Treating: The completed investment of wastewater treatment facility is preferred as end-of-pipe sector of tanning industry. Wastewater treatment standard must be critically followed up to the environmental friendly standards to prevent environmental affection from the untreated waste.
6. Dumping: At the bottom of the hierarchy, dumpling ithe leastss preferable option due to tanning wastes that cause high environmental pollution and furer affectto human health. In fact, due to the costs of treating tanning waste, many tanning manufacturers rid ofdeal tanith illegyactivities.

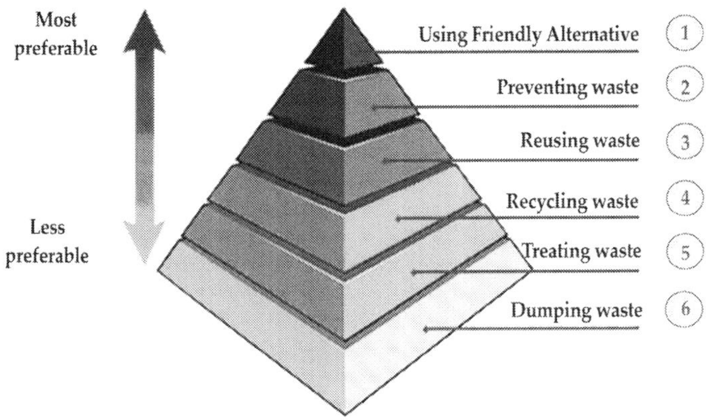

Source: https://www.intechopen.com/books/management-of-hazardous-wastes/

ENVIRONMENTAL IMPACT OF TANNING

One tone of hide or skin generally leads to the production of 20 to 80 m3 of turbid and foul-smelling wastewater, including chromium levels of 100–400 mg/L, sulfide levels of 200–800 mg/L and high levels of fat and other solid wastes, as well as notable pathogen contamination. Pesticides are also often added for hide conservation during transport. With solid wastes representing up to 70% of the wet weight of the original hides, the tanning process comes at a considerable strain on water treatment installations.

Tanning is especially polluting in countries where environmental norms are lax, such a in India,- the world's 3d largest producer and exporter of leather. To give an example of an efficient pollution prevention system, chromium loads per produced tone are generally abated from 8 kg to 1.5 kg.

Raising animals for food and leather requires huge amounts of feed, pastureland, water, and fossil fuels. Animals on factory farms produce 130 times as much excrement as the entire human population, without the benefit of waste treatment plants. The U.S. Environmental Protection Agency (EPA) has even acknowledged that livestock pollution is the greatest threat to our waterways.

Although some leathermakers deceptively tout their products as "eco-friendly," turning skin into leather also requires massive amounts of energy and dangerous chemicals, including mineral salts, formaldehyde, coal-tar derivatives, and various oils, dyes, and finishes, some of them cyanide-based. Most leather produced in the U.S. is chrome-tanned; all wastes containing chromium are considered hazardous by the EPA.

Tannery effluent contains large amounts of pollutants, such as salt, lime sludge, sulfides, and acids. The process of tanning stabilizes the collagen or protein fibers in skins so that they actually stop biodegrading—otherwise the leather would rot right off your feet.

People who work in and live near tanneries suffer too. Many die from cancer possibly caused by exposure to toxic chemicals used to process and dye the leather. The Centers for Disease Control and Prevention found that the incidence of leukemia among residents in an area near one tannery in Kentucky was five times the U.S. average.

Arsenic, a common tannery chemical, has long been associated with lung cancer in workers who are exposed to it on a regular basis. Studies of leather-tannery workers in Sweden and Italy found cancer risks "between 20% and 50% above [those] expected."

Additionally, raising the animals whose skin eventually becomes leather requires vast quantities of water and wide tracts of pastureland, which must be cleared of trees. Runoff from feedlots and dairy farms also creates a major source of water pollution. Huge amounts of fossil fuels are consumed in livestock production as well; by contrast, plastic wearable's account for only a fraction of the petroleum used in the U.S.

pollution load decrease achievable according to the united nations development program posts precise data on the abatement achievable through industrially proven low-waste advanced methods, while noting that even though the chrome pollution load can be decreased by 94% on introducing advanced technologies, the minimum residual load 0.15 kg/t raw hide can still cause difficulties when using landfills and composting sludge from wastewater treatment on account of the regulations currently in force in some countries.

One of the major environmental problems, causing a serious threaten over the world, is the contamination of atmosphere components with toxic chemicals. Unfortunately, the most xenobiotic compounds, produced annually at huge amounts, are persistent in the environment and have carcinogenic and/or mutanogenic effects. The main sources of xenobiotics are wastewater, landfill leachates, and solid wastes. Xenobiotics include phenols, plastics, hydrocarbons, paints, dyes, pesticides, insecticides, paper and pulp mills, pharmaceuticals, and others.

Biological processes, being eco-friendly and cost cheap techniques, were proposed for xenobiotic degradation to overcome these problems. White-rot fungi, especially the basidiomycetes, are the most tolerant microorganisms to high concentrations of pollutants, giving their exceptional abilities for biodegradation in aqueous environments and soil and have been investigated extensively for their bioremediation capacities. Fungal bioremediation is a promising tool since the metabolic potential of such microorganisms converts most of the environmental pollutants to less hazardous or non-hazardous compounds with less input of energy and time. White-rot fungi are the unique organisms that show the capacities of degrading highly toxic organics and recalcitrant compounds.

The key enzymes of their metabolism are extracellular ligninolytic enzymes that enable fungi to tolerate high concentrations of toxic substrates. These enzymes have potential applications in a large number of fields, including the chemical, fuel, food, agricultural, paper, textile, and cosmetic industrial sectors. Their capacities to remove xenobiotic substances and to produce others, which are less or non-toxic, make them a useful tool for bioremediation purposes.

The potential of white-rot fungi can be harnessed thanks to emerging knowledge of the physiology and morphology of these microorganisms. This knowledge could be transformed into reliable and robust waste treatment processes. The importance of high extracellular levels of these enzymes to enable the efficient degradation of xenobiotic compounds under in vivo conditions relates to the sorption and complication of enzymes in soil and the probable loss of much of their activity once externalized.

ORIGIN, THREAT, AND BIODEGRADATION OF XENOBIOTICS: THE ENZYMATIC SYSTEM INVOLVED IN XENOBIOTICIODEGRADATION.

90% of all organo-pollutants produced every year are not properly disposed. These pollutants persist in the environment long after use. Some primary sources of xenobiotics include; wastewater, leaches from landfills and solid residual wastes from industries. Volatile organic compounds (VOCs) are incorporated into the structure of plastics foams, packaging, solvents, paints, adhesives, and floor and wall coverings.

Domestic and industrial wastewater also contains phenolic compounds, hydrocarbons, dyes, and pesticides. Surface runoff water leaches a significant amount of these harmful substances from landfills. The landfill leach contains high levels of ammonia and nitrogen. In addition, phenols and hydrocarbons are also leached from landfills. Combined effects of this toxicants and their physiochemical properties is responsible high toxic levels. Their harmful constituents remain after multiple treatments. Fungus strains can remedy the pollution causes by landfill leaches. Some fungus strains include the Trametes trogiie, P. chrysosporium, and the Lentinus tigrinus. Through their cellular enzymes, they reduced organic pollutants such as phenols, hydrocarbons, and oxygen demanding wastes. They also offset the toxicity of two folds diluted LFL (laaches). However, However, these strains of fungus are sensitive to high levels of toxic compounds. Phenols and phthalates significantly reduce the growth and amount of cellular enzyme secretion from fungi. Other strains such as autochthonous and allochthonous were noted by Tigini to be similarly effective in treatment of land fill leaches. In addition, the strains showed a broad spectrum of action.

Examples of domestic solid waste residues are foodstuffs, paper, and garden wastes. Solid waste are also generated during street cleaning, trimming of trees and garbage collection in public litter bins. Wastes from commercial and industrial activities contains a higher percentage of metals and plastics. Huge quantities of solid wastes are also generated during construction and demolitions. All these wastes constitutes urban waste, which contains a high proportion of xenobiotic compounds. This wastes exists as landfill leaches and the more volatile constituents are in the gaseous form. Its major compounds are dimethyl phthalate phenol, benzoate and phthalic acid.

Phenolic compounds are widely found in nature. They are concentrated in plants and aquatic life. Microorganisms are capable of degrading them into constituent compounds. Phenols are highly toxic at little concentrations. Processes that produce these wastes include; coal refining, paper manufacturing, petrochemical refining, drug making and phenol manufacturing. The phenol molecule is an essential component of plastics, pesticides, detergents, textiles, explosives and drugs. Therefore, it's critical in the manufacture of synthetic resins. The adverse effect of phenols felt in environment includes the thinning of the ozone layer. To humans it's a dangerous carcinogen. Use of natural methods to degrade phenols is the only effective method of reducing its effects. The biodegradation process is energy efficient and produces less toxic metabolites. The noted strains of fungus that naturally breakdown phenols include: Trametes villosa, Lentinus eodes and Trametes versicolor.

Chlorinated phenols are dangerous pollutants degraded by fungi. Likewise enzymatic action breaks down the toxic pollutants to less toxic chemicals. According to Ehlers and Rose restrained WRGeduce chroninatedchlorinated phenolic compounds to less harmful constituents.

Plastic contains toxic hydrocarbon constituents such as polystyrene, polyvinyl chlnd other harmufulharmful polymers. The natural process of degrading plastics using natural microorganisms can take hundreds of years. This is due to complex polymer structures made of longs chains of atoms. Biodegradation of plastics is still an ongoing field of study. An efficient method is yet to be developed that doesn't produce more harmful components that contribute to further ecological damage. The P. chrysosporium was found to degrade plastics like nylon.

Another class of harmful pollutants are polycyclic aromatic hydrocarbons. These are constitutes of effluent waste from the petrochemical production process. Treatment of polycyclic is rather expensive and further degrades the environment. Biodegradation offers a better solution since its cost effective and efficient. This involves use of microorganisms to restore sites contaminated by various types of petroleum baatives. However, However, aromatic ring hydrocarbons are difficult to break down because of saturation. According to Olusola and Anslem, the Pleurotus pulmonaroius can break down crude oil. Other fungus strains such as the P.chrysosporium and Pleurotus can remedy soil contaminated by anthracene and pyrene.

Paint contains volatile organic compounds and additives such as lead released to the environment during flacking. Constituents in paint such as emulsifiers and bases are remedied through chemicals and microorganisms. Various strains of fungus successfulpaints. However, However, the growth of fungi on painted surfaces often leads to undesirable effects such mechanical and appearance damage.

Textile industries are accountable for contributing significant amounts of pollution. Dyes used in the production process are often discharged directly into the environment. These dyes eventual evaporate and break down forming products that are toxic. Fungi through their unique enzymatic degradation can break down these dyes. Ongoing research is focused on understanding the mechanism better and trying to identify toxicity levels of the final products. The mechanism involves a series of steps producing intermediate compounds. So far, fungal isolates and enzymes which de-color dyes such as the blue dextran, cresol red and Poly R48 have been discovered. Ben Younes noted that laccase from the thermophilic Scytalidium thermophilum speeds up the de-colorization of the Congo red and triarylmethane dyes, frequently found in effluents.

Effluents discharged during paper manufacturing contain high concentrations chlorinated organic compounds used in the bleaching of pulp. This effluent also contains organic halides and phenols. Most of the bleaching agents are chlorinated making them highly toxic, in addition to being mutagenic. Treatment of these wastes is through the anaerobic process carried out in low oxygen concentrations. Fungal enzymes laccase from Coriolopsis gallica successfully de-color used dyes, which forms a major portion of the effluent. These fungal species also directly removes chloro-phenols from bleach effluents. Fungus laccases function to improve the quality of the pulp.

Most herbicides are organophoshorous compounds. When applied in large quantities they lead to extensive contamination. They persist in the environment for long periods after use, due their slow rate of degradation.

5 THE DETECTION AND CONTROL OF HAZARDOUS WASTE SOURCES

INTRODUCTION

The definition of the word 'hazardous' differs from one country to the other. Generic definition of hazardous could refer to the waste material or a combination of the waste material that pose a considerable threat to the environment and the living beings, and this definition remains the same for both developing and the developed countries. The majority of the toxic/ hazardous waste is generated while some products are being manufactured for domestic use and and/ or also for some further industrial use.

The management and the risk assessment of the hazardous waste could only be efficiently done when the hazardous waste is properly detected/ identified, so that the required actions could be taken, depending on the nature of the waste material. As hazardous waste materials are produced in huge chunks, irrespective of the fact that a country is a developed or a developing one. The control mechanism and effective management of the waste as always is dependent on the timely/ proper detection and then applying the streamlined procedures which are in line with both the local and the international standards of the waste management. The control mechanism of the overall management of the hazardous waste material becomes functional in the most productive manner when the sources of the hazardous waste material produced are identified. This would also help in minimizing the waste material by enhancing the manufacturing progress so that minimal waste is produced and also it is processed in the best possible manner. http://www.eolss.net/sample-chapters/c09/e1-08-00-00.pdf

IDENTIFICATION PROCESS OF THE HAZARDOUS WASTE MATERIALS

Proper identification of the hazardous waste material coming through any source is essential in order to succeed in the management program for the hazardous waste materials. According to the regulations used in several developed countries, the requirements is that any person or a facility that generates or produces any form of the waste material, should be investigated, if it contains hazardous material or not. The following are the questions that should be answered for the sake of identification of the waste material:

- Is the waste material produced, a 'solid waste'?
- Is the waste material produced, is the listed type of the 'hazardous waste'?
- Does the waste material produced exhibit a certain characteristic of the hazardous/ toxic waste material?

The following are some of the key steps that are considered in the identification of the waste material produced:

- **Is waste material produced is of solid type?** – The identification of the hazardous waste material initiates with a very obvious point, which is that, in order for any material to be categorized as a hazardous type of a waste material, in the first place, it has to be a 'waste' in the first place. But, in order to decide whether a material is a waste or not, is not a very simple or easy process. As an example, a particular material which one person or any institute/ facility/ site rds might seemsseem to be a great value to another person or an institute/ facility/ site. Such a material could possibly be an aluminum can which might be disposed from one site but could be of use for another. There are a set of rules developed in the developed countries in order to assist in the determination of a material to be waste or not. According to Resource Conservation and recovery Act (RCRA), the 'solid waste' term is used in place of just 'waste' which is considered to be a much common term. Under the RCRA, solid waste refers to any waste material which could be a semisolid, solid or a liquid.

- **Is the waste material excluded?** – Only a very small percentage of the solid waste materials under the RCRA qualifies as the hazardous waste material. At first glance, one might think that the distinguishing between the non-hazardous and the hazardous types of the waste materials will be a quite simple method that include the toxicological and the chemicis only. However, However, that is not the case as several other factors must also be considered, in order to evaluate that what actual hazard does the waste material possesses. The regulations imposed for a particular waste material types might not be that practical, fair or may be not desirable otherwise, which is again regardless of the type of hazard the waste contains. As an example, the domestic waste material could contain certain types of the dangerous chemicals such as the pesticides and solvents etc., but the making of the domestics subject to the strict form of RCRA regulations for the waste management might lead in creating a variety of practical problems and concerns. The EPA and Congress exempted or the certain types of the excluded waste materials, such as the domestic waste materials that comes through the definition of the hazardous waste material regulations. The determination of the fact that a certain type of a hazardous waste material type is exempted or excluded from the hazardous waste material regulation is actually the 2nd step in the identification process of the RCRA hazardous wastes. Just after the determination of a solid waste material that it is not excluded (somehow) from the regulations of hazardous waste materials, then the analysis is further continued to evaluate that what actual type of a chemical hazard is there which the waste material under test possesses.

- **Does the waste material exhibit a particular characteristic**

or is it a listed hazardous waste material? – The final steps towards the identification of the hazardous waste material determines the fact that whether the waste material contains chemical in sufficient amount to pose a physical hazard to merit the regulation. The steps involved in the identification process of the hazardous waste material include the evaluation of the waste material in the light of the definition of regulation of the hazardous waste material.

https://www.epa.gov/sites/production/files/2015-09/documents/hwid05.pdf

THE SOURCES OF THE HAZARDOUS WASTE MATERIAL PRODUCED

The hazardous waste term very often include the likes of the by-products coming from the domestic, healthcare, industrial and the commercial facilities. Rapid development and the enhancement of multiple types of the industrial technologies and the practices might increase the generation of the hazardous waste materials. The majority of the toxic waste materials are generated during the manufacturing phase of the products used in the consumption or in further application in the industries. All of the sources of the hazardous waste materials, which are known to discharge gas, liquid or the solid wastes could be regarded as the hazardous waste sources. The following is a brief account of the different sources of the hazardous waste materials:

- **Agricultural land/ agro-industry** – The hazardous waste materials coming from the agricultural land and/ or the agro-industry could expose the human beings to fertilizers, pesticides and several other types of the toxic veterinary product waste materials. Farms are considered to be the prime source of such wastes being produced and then the agrochemicals could leach in the environment, while they are in storage or they could also harm after their actual application.
- **Building materials** – The roofs and he pipes that are made by employing the materials that make use of the materials like the copper, asbestos or several other types of the materials might

give rise to the source of toxic waste material.

- **Commercial waste material** – The waste sources through commercial means include the automobile repair workshops, gas stations, and dry cleaners etc. The type of the toxic wastes produced by such sources is dependent on the types of services they provide.
- **Contaminated sites** – Such sites are contaminated with the toxic waste materials which are usually the result of those activities, which either uses or produces the toxic substances or it may also be due to some accidental spills of the material. The former sites of the industries, which make use of such hazardous waste materials are also known to belong to this group.
- **Domestic** – The household stock which include the different types of the toxic substances such as the dry cells, batteries, wood preservatives, paint thinners, furniture polishes, stain removers, rat poisons, paints, mosquito repellents, pesticides and herbicides, fuels and the disinfectants and several other types of the automotive products.
- **Health care facilities** – Such facilities are known to be the sources of the contaminated needles, blood and the pathological waste materials. Some particular sources of such type of the waste materials include veterinary clinics, dentists, blood banks, morticians, home health care clinics, hospitals and the medical laboratories.
- **Institutional sites** – The institutional toxic waste material sources are usually the research centers, research laboratories and the military installation sites. Some of the military installations are used for the production and storage of the ammunition and are also used for the purpose of a testing ground for the hardware (military). The military establishment is also known for carrying the activities, which would generate several other types of the toxic waste materials of the commercial, domestic and the industrial nature.
- **Industrial sites** – The hazardous waste material is produced

through a range of industrial activities, such as, the waste material coming from the fuel (petroleum) industry which include the refinery products, for example, tar, sludge drained form the tanks (storage), the impurities of the cyanides and phenol and those are all usually the part of the waste stream.

- **Mines/ mineral processing sites** – Such facilities are known to handle various types of the hazardous substances, which are present in the products, additives and also in the waste material produced.
- **Solid waste material disposal sites** – Such sites are usually the disposal sites for the wastes coming from the municipal sites and these also include the toxic wastes, which have not been separated properly from the other non-toxic types of the waste materials. The solid waste material disposal sites are considered to the prime source of the pollutant laden leachate in the third world countries especially.

SOME USEFUL METHODS FOR THE HAZARDOUS WASTE MANAGEMENT

The hazardous or the toxic types of the waste material shave become a very critical parte environmental and human beingsbeings' health issues which raises several serious types of concerns in multiple countries worldwide. In the modern setting of the hazardous waste material management, the following 4 strategies have been adopted to get the best possible results:

1. Minimizing the quantity of the waste material produced
2. The recycling process of the industrial waste material type
3. The treatment process of the hazardous waste material
4. The collection/ storage, transportation and disposal of the hazardous waste material in a very environmentally sound way

All the 4 types of the strategies mentioned above are quite crucial and practical and are found not be exclusive of one and other. When a person/ facility/ site/ institute deals with a particular type of a hazardous waste problem then there might be a need to possibly employ a combination of the 4 approaches mentioned above. The following is a brief discussion of all these 4 strategies for the hazardous waste management:

1. **Minimizing the quantity of the waste material produced** – Reduction in the quantity of the waste material produced is the top most priority in this process. There are three prime waste material reduction schemes in practice which are described briefly as follows:

    - **Process modification** – Very often, the employed industrial process could be modified in a way so that the usage of the raw materials in the process is first optimized, which then result in the reduction of the hazardous waste material to the bare minimum level. As an example, in case of the zinc electroplating, the substitution is made for the sulphate salt with the chloride compound, by using the slight modification technique in the process. This eventually leads to the elimination of the cyanide issue.

    - **Waste segregation** – The segregating of the waste material (hazardous) streams from that of the waste material (non-hazardous) stream result in the reduction of the hazardous waste material volumes, which makes the waste material, very easy to treat, eventually.

    - **Waste Concentration** – The waste material generated is capable of being concentrated by employing the evaporation decantation or the precipitation techniques, which in a way means that the waste material volume could then be reduced significantly by using such methods. Incineration technique is very often used in order to reduce the net waste material volume that is to be handled. It is a very useful technique for handling the waste material disposal, but, it is also a fact that the total cost incurred in the process is known to exceed the net gains usually.

2. **The recycling process of the industrial waste material type** – A lot of the substances found in the waste materials have some good value to be used again. Such things include the likes of wood

fiber obtained from the paper products, glass material and metal etc. Numerous scientists now have developed different ways for the purpose of recycling a range of waste materials, in a way so that they could be used again. Almost all of the waste materials produced are capable of being recycled. But, for some of the materials, there is a case of possibly more energy to be expended in the recovery process as compared to the achieved recovered value that is being warranted. There are basically two general ways for the processing of the hazardous waste materials which are explained briefly, as follows:

- **Waste Reuse** – In some of the known cases, the waste material produced could be employed as a raw material, by applying very little processing over it. The transfer of the waste material without any reprocessing to another site/ facility is known as 'waste exchange' or 'waste reuse'. The unwanted type of the waste materials such as the untested waste materials or the outdated chemicals could be reused without applying any formal reprocessing techniques over it, which are not meeting the high standards of quality control in the purchasing industry. Then there are certain types of the process waste materials as well, such as, the cardboard used for making the pulp (paper), oils that could be used as a source of energy (fuel) and the copper or other types of the metallic salt solutions which are used for the process of metal recovery. This list also includes a lot of other types of the waste materials, which could be reused as the industrial feed stocks.

- **Waste recycling** – The recycling of the waste materials produced actually differs from the reuse, in such a way, so that the waste material should first be processed before it could be used for any other process such as, manufacturing. When a transfer of waste without doing any processing or treatment is not possible then the reprocessing of that waste material produced is known as recycling. As an example, the baghouse dust that is known to be coming out of the scrap steel processors, which may contain nearly 25% of the zinc oxide, could be contaminated with the sulphuric acid waste material in order to make the pickle acid (galvanizer). The spent liquor of pickle which is known to contain some 8 to 10 % of the zinc sulphate and a few iron salts, is usable then, as a mean of fertilizers that are found in the agricultural fields. The use of any organic solvents waste material is considered as the top example for the recycling waste process

3. **The treatment process of the hazardous waste material** – After the process of the waste material recovery, the water (waste) that contains the toxic chemicals (waste) should then be detoxified and also neutralized by making them go through the required treatment for the purpose. There are plenty of different types of available technologies as well for the purpose of treating thezardous materials, before they get disposed offof permanently. The aim of such technologies use dis to modify the chemical and/ or the physical properties of the waste materials so that they would not be harmful anymore. The selection of a particular technology for the processing of the waste materials depends upon a number of factors such as, the desired characteristics of the stream (output), nature of the waste material and finally the energy and the economic considerations. The technologies employed for the treatment processes could be categorized into the following main groups:

    - Chemical treatment
    - Physical treatment
    - Biological treatment
    - Incineration
    - Solidification

The following is a brief description of the above mentioned technologies:

a. **Chemical treatment** – The chemical treatment is employed in order to facilitate the complete breakdown associated with the toxic waste materials and even more, usually for the purpose of modification of the chemical properties of the waste materials, for example, for decreasing the water solubility or may be to neutralize the alkalinity or the acidity contained within it. The techniques used involves the likes of, oxidation, neutralization, chemical reduction, oil/ water separation, fuel/ solvent recovery and heavy metal precipitation.

b. **Physical treatment** – The physical treatment is conducted by employing various types of methods such as the phase separation to be the top most used method. It include 3 steps, namely, 1. Lagooning, 2. Protracted storage in the storage tanks and, 3. Sludge drying in the beds. The Lagooning and the protracted tank storage steps usually used in tandem in order to separate the particulate impurities.

c. **Biological treatment** – The gross impurities that are collected/ obtained through the sewage treatment are known as sludge, collectively. The process as a whole is known as sludge processing which has become very critical since the enhancements are done in the industrial waste water treatment. Some of the most typical technologies for the sludge processing include the likes of digestion, conditioning, dewatering/ thickening, composting and solidification. The following is a brief explanation of these typical technologies mentioned:

    - **Conditioning** – In this one, the sludge is first of all exposed for a stipulated period of time to the open atmosphere. This exposure continues on, until the consistency required is achieved.

    - **Digestion** – Here, the sludge is process with bacteria where the breakdown happens for the turning of the long chain compounds into much simpler ones.

- **Composting –** In this step of the process, the organic matter, which is present in the waste material sludge is then converted into usable form of a material (stable).

d. **Incineration** – Incineration is a kind of a process which leads to the burning of the waste materials produced. This is also another technique which a large number of industries and several large cities of the developed countries use today, in case there is no presence of large size vacant areas within the vicinity of their cities. The majority of the toxic waste materials used are actually detoxified by using the process under discussion. This method is also known very well for its capability of performing the waste minimizate detoxification and waste disposal. However, However, the process is not too practical in many parts of the world, which is due to its excessive cost of operation, especially in the case, when the waste material's heat content is not actually reutilized.

e. **Solidification** – The solidification process actually work on to convert the liquid form of the waste material into some rock hard and an insoluble material and are then employed as a kind of a pre-processing which is done before the landfill disposal. This is done by actually mixing the produced waste material with multiple other reactants in order to generate a very much solid mass. The basic aim behind this process is to immobilize the toxic components of the waste material, so that such components could not actually leach out at the landfill site (disposal).

4. **The collection/ storage, transportation and disposal of the hazardous waste material in a very environmentally sound way –** In relevance to the above discussion, the current section should be broken down into the following parts for easier understanding of the complete process:

- **Storage, transportation and dis** – The disposal of the waste material might seemsseem to be a straightforward activity, but in reality it is quite difficult and the process involves multiple levels. The different levels or stages for the disposal include the collection, interim storage, transportation, processing/ treatment and disposal. The phases mentioned here are known to be very interdependent, both in the ways of their organization and the technique. The protected collection and the transportation of the toxic waste materials from a critical link in the chain right between the generation points and its place of processing and disposal. In many ways, same kind of precautions actually apply to the toxic waste materials in transit as it applies to the hazardous goods carriage. Still, some more problems do get trigger from the toxic nature of the certain waste materials because of the reasons given below:

    i. In general, the waste material is known to have no perceptible value (economic) to the generator.

    ii. The physical and the chemical properties of the generated waste materials might not be known precisely, because of this being a very complex mixture, from which the useful (economically) components have been extracted.

    iii. The mixing of the non-compatible type of the waste materials for the sake of convenience in the transit could result in an acute hazard, with occurs either spontaneously or on the processing or disposal, for example, the ether waste material mixture which is known to contain a residue of sodium with an aqueous type of ether was material would explode.

Therefore, for the reason given above, in order to have a protected and a secure disposal system for the toxic waste materials, there is a need to have a proper means of collection, transportation and the storage system. The waste materials which are of non-compatible type should first be properly segregated and then they should be separately transported.

- **The disposal of the hazardous waste material** – The ultimate disposal of the toxic/ hazardous waste material also require that they should be planned very carefully. The following are the 4 different types of wastes through which the hazardous waste materials could be disposed finally:

    i. **Landfill disposal** – The disposal of the toxic waste material using such method is quite popular in multiple countries. Landfilling refers to the process of storing or disposing the harmful toxic substance under the ground.

    ii. **Incineration** – The technique used in incineration is to burn the toxic material for the sake of their disposal. It is also a popular method used in multiple countries and within many industries worldwide. The majority of the toxic materials are known to be detoxified in this useful process. The operational cost for this process could be very high, especially when there is no reutilization of the waste material's heat content.

    iii. **Dumping at sea** – it is a disposal technique which was useful in the pts convenience and inexpensive operations. However, However, the use is not much preferred as it affects the marine life and the sea food which human beings consume. Waste Disposal Methods. 2017. *Waste Disposal Methods*. [ONLINE] Available at: https://www.gdrc.org/uem/waste/disposal.html

    iv. **Underground disposal** – The underground disposal system is desirable especially in the developing countries where the over ground installation could be impossible, difficult, not much profitable or undesirable for the deterioration of the environment. http://www.iswa.org/uploads/tx_iswaknowledgebase/P13-Kaliampakos.pdf

CONCLUSION

As we learned above that there could be several sources of the hazardous waste material that include industry, agricultural land, domestic consumption leftovers, commercial waste, mines/ mineral sites, institutional wastes, healthcare sites wastes, solid wastes and building material wastes etc. There are certain factors involved, which should also be analyzed as, those factors results in the production and also mishandling of the toxic waste material. Once the sources are derived then their identification should be done so that the most effective hazardous waste management plan could be adopted. Finally, any plan for the handling or the overall management of the hazardous waste material is adopted then that has to be compliant with the rules and regulations of the waste material authorities and for the third world countries that must also follow the standards used in the developed countries.

6 HAZARDOUS SOLID WASTE DISPOSAL SITES

INTRODUCTION

The management for the hazardous waste materials in any society is a very daunting challenge to accomplish for the health of the human beings and the safety of the environment. It becomes an even more complex job in the developing countries due to lack of waste management infrastructure and several other dependencies as well. The solid waste as opposed to the liquid waste material is way difficult to dispose as several complicated processes are required to do the needful. The job for its completion requires very sophisticated waste disposal sites which are equipped with different facilities/ equipment, based on the waste disposal method being used. Such facilities ensures proper disposal of the hazardous waste material and also ascertain that no toxic wastes are left for making the environment contaminated. Recycling of the waste materials for reusability or at best, the prevention of waste materials being produced are the best ways to counter pollution due to hazardous waste materials, However, there are some waste types which are not at all possible to be disposeffof in the manner so that they could be recycled or prevented from existence in the place. In such cases, the waste disposal sites becomes effective for their proper disposal.

UNDERSTANDING THE SOLID TYPES OF THE WASTE MATERIALS

Distinguishing between solid, liquid, gaseous or any other type of wastes is very important in the waste management industry as only then there could be some waste management programs adopted as per the relevance of the waste material. The solid waste could be any refuse, garbage or sludge from a water supply plant (treatment), or from air pollution control facility and several other types of the discarded materials which could be of solid, semi-solid or even of liquid type. The solid wastes may result from the garbage of household refuse or from different types of industries, commercial facilities, hospitals, clinics, community activities, mines and the agricultural activities. The solid materials does not include anything like the solid or even the dissolved waste materials present in the domestic sewage or the same, present in the irrigation flows or different types of the industrial discharge. In the simplest terms, the solid waste materials are the discarded, refused or the abandoned types of the waste materials that could range from solid, semi-solid, liquid or of the gaseous type. Some of the most commonly found waste materials of solid type in both the developing and the developed countries include the following:

- Sewage
- Wasted tires
- Latex type paints
- Scrap metal
- Garbage/ trash
- Furniture
- Toys
- Vehicles
- Appliances
- Oil
- Anti-freeze material
- Paint cans
- Compressed gas cylinders
- Empty aerosol cans

- Demolition or the construction debris
- Asbestos
- Etc. etc.

Any solid waste material in the waste management industry is considered to be discarded, in case where that is being abandoned by any of the following:

- Being disposed off
- Being incinerated or burned, which also includes being burned in the form of a fuel for the sake of recovering the useful energy (usable)
- Being collected, stored or treated chemically, physically or biologically (apart from being incinerated or burned) instead of or may be before the waste material is being posed offof.

Any of the solid material getssposed offof when it is either, discharged, dumped, injected, deposited, leaked, spilled or placed into any facility on land or water . This is done for the reason that any of such material or may be its constituents may not add to the environment or gets emitted in the air or there is also a possibility of being discharged into the surface or the ground water. What is Solid Waste - NYS Dept. of Environmental Conservation? 2017. What is Solid Waste - NYS Dept. of Environmental Conservation? [ONLINE] Available at: http://www.dec.ny.gov/chemical/8732.html

REASONS FOR THE SOLID WASTE SITES/ FACILITIES TO EXIST

Everything is needed for a reason. When we trace back the necessity of the solid waste material disposal sites then a number of reasons for it could be seen which are given below:

- **Human beings produces a lot of waste materials** – Human beings have a lot of consumption in their daily routine life which could be of household, commercial or industrial nature. According to the numbers given by US Environmental Protection Agency (EPA), the citizens of US are known to produce the trash or waste materials in excess of 250 million tons every year. The recycling or composting done for the waste material produced is just about 1.51 pounds out of the total 4.4 pounds of the waste material which an individual in US generates per day. The best acthuman beingsbeings' part here could be to reduce waste material by following the 4 R's process as follows:

- **Reduce –** Be careful with the environment and try to reduce as much waste material as possible.
- **Re-use** – If there is any possibility for a consumable item to be reused then make sure that you go for it in order to contribute far less to the waste material being produced.
- **Recycle** – Recycling helps the environment with the waste material being recycled in order to bring it back in our use in one way or the other. This minimizes the waste material and makes it a kind of a productive thing which is processed and becomes reusable again. The government provides facilities for dumping the identified recyclable products in a particular place which could then be recycled in order to bring it back in our use**e-purposeing** – Repurposing is the act of rechanneling an item in order to bring it in an unintended use. This again could bring in significant impact in the reduction or the management of the waste material.

- **Human beings produce toxic waste material –** Waste of any kind if it could not be recycled can be of significant concern for the residents being impacted with that in thy. However, However, the toxic waste material which contains hazardous waste and that is dangerous for the health of the human beings and for the environment, at large. The health implications of some of the chemicals, such as, the Biphenyl- (BPA) is very much clear at this point, but still this one is very commonly found in the plastic products, including toys and are very poorly regulated. The unprecedented type of the toxicity present in the trash/ garbage worsen the issue and such waste materials needs to be taken as soon as possible for its proper disposal. Unfortunately, in most of the developing countries, there is no known solution for dealing with such types of the waste material, effectively.

- **Regulations known toavoring some special interests** – The waste material processing/ handling of late has become a very much commercial and a money making thing. This makes the residents outmatched by the industry lobbyists at the state. The regulations, currently are making things very daunting for the states or the communities to regulate effectively, the waste management facilities. On top of that, devoting to the resources for the waste reduction or the recycling programs also becomes difficult. The waste industry of today in both the developed and the developing countries. Some of the large corporation for handling waste management are dominating some good aspects of the waste material market and they make huge profits by operating at the landfills and/ or the incinerators, and that also include the recycling facilities as well. Since, the waste management industry has fostered into a very big business, the corporate need to make a profit which actually outweighs all the needs of the community of reducing wastes, providing recycling facilities and above all protecting the environment. Furthermore, the need of such kind could only be catered when there are excellent facilities or the waste disposal sites available for the processing of the produced waste material. The regulations needs to be imposed nationally as one program running very effectively in one state could be affected with a poor program running in the next state as the toxicity in the environment, land and sea could penetrate into the vicinity of the neighboring state. The government at all this faces huge challenges in maintaining such standards country-wide. The government is continuing to work in this regard hand-in-hand with the officials in the waste management industry in US in order to allow for considerable expansions of the landfills, enhanced tonnage of the waste in the incineration and also in developing some of the new facilities, such as, the trash transfer stations. Government in most of the countries feels very pressurized in the urge of finding the places for waste material disposal as very few resources are devoted for working on the development of the recycling functional programs. Instead, the government finds it more suitable to look for quick and short-sighted fixes or solutions in waste management. Toxics Action Center. 2017. *Toxics*

Action Center Waste | Toxics Action Center. [ONLINE] Available at: https://toxicsaction.org/issues/waste/

THE PLACES/ FACILITIES IN USE FOR THE DISPOSAL OF HAZARDOUS WASTES

The hazardous/ toxic types of the waste materials are sent to the waste disposal sites/ facilities for the storage, treatment or their disposal. The waste disposal sites/ facilities are also sometimes referred to as the treatment, storage and/ or the disposal treatment sites/ facilities. The activities of such facilities could be seen below:

- **Treatment** – The different types of processes, such as oxidation or incineration are used for altering the composition or character of the toxic waste materials. Few of the processes used for treatment enable the waste material in order to be recovered first and then reused in different types of the manufacturing settings. On the other side, there are some other types of treatment processes that work on to dramatically reduce the hazardous waste amount.

- **Storage** – Temporarily accumulating the toxic waste material until they are taken for ultimate treatment or disposal. The toxic type of the waste material is usually stored before any treatment or disposal is done and they must be stored in tanks, containers, drip pads, containment buildings, waste piles or may be the surface impoundments which are known to be compliant with the Resource Conservation and Recovery Act (RCRA) in USA.

- **Disposal** – The permanently containing toxic wastes are taken for disposal. The most commonly used facility for the disposal of the waste material is a landfill where the toxic waste materials are disposed in the most carefully manufactured units that are designed for protecting the surface water and the ground water resources.

For the purpose of different types of solid waste material disposal, the following are the facilities/ place that are used for doing the needful:

- **Containment Buildings** – The containment buildings are known to be enclosed completely and they have the structures, which are self-supporting. It means that such a building has 4 walls, floor and a roof. The place is used for the sake of storing and also treating the hazardous waste material (non-containerized).

- **Landfills** – The landfills are the engineered or the excavated sites where the solid waste material type is deposited for the recovery/ recycling or for its ultimate disposal. The units used are designed in such a way so that there is minimum chance for the release of the hazardous waste material into the environment. The design standards used for the toxic waste material landfills requires the following:
    - Double liner
    - Leak detection system
    - Runoff, run on and the controls for the wind dispersal system
    - Double collection (leachate) and the associated systems for the removal needs
    - The quality assurance program for the construction

The operators are also required in such sites to comply fully with the monitoring, inspection and the with the release response requirements. As the landfills are known to be the permanent type of the disposal sites and they are also known to be enclosed with the placed waste material, so they are subject to the closure and for the post-closure maintenance requirements which include the following:

- Installation and then maintenance of a final cover
- The continuous operation required for the leachate accumulation and the removal system there is no longer detection of the leachate system
- The monitoring and/ or maintenance of the system for leak detection
- Maintaining the system for the monitoring of the ground water

- The prevention of the storm water runoff and run on
- The installation and the protection of the benchmarks (surveyed)

- **Incinerators** – The incinerators are those kind of enclosed devices which makes use of the flame combustion (controlled) for the purpose of thermal treatment of the produced and/ or accumulated type of the toxic waste material. When the process is properly performed then the toxic type of the organic constituents contained inside the hazardous type of the waste materials are completely destroyed. On top of that, the total volume of the waste material is also considerably reduced which makes the disposal process much easier.
- **Land treatment units** – The land treatment units make use of the soil microbes (naturally occurring) and the sunlight for the treatment of the hazardous waste material. This is accomplished with the application of the toxic waste material directly into the soil surface or may be associating with the upper layers of the soil for the purpose of transforming, degrading or immobilizing the toxic constituents. The treatment units (land) are dependent on the chemical, physical and the biological processes that occur in the topsoil layers for containing the waste. It is because of this that the land treatment units are not supposed to have the liner system, or the leachate removal or the collection system. Before, any form of the toxic waste material could be placed inside the land treatment unit, the operators are supposed to mandatorily complete a treatment demonstration for demonstrating the ability and the effectiveness of the unit for the purpose of treating the toxic waste material. Once, the unit becomes operational then the operators are required to mandatorily monitor it (the unsaturated zone monitoring) for ensuring that all the toxic constituents are adequately treated. The unit closure consist primarily of putting a cover (vegetative) above the unit and certifying that the toxic type of the constituent level do not go beyond the background levels when in the treatment zone.

- **Surface Impoundments** – The natural topographic depressions and the surface impoundments, excavations (man-made) or the diked vicinities which are primarily made from the earthen materials (still lined with the materials which are man-made) are used to store, treat and/ or dispose the hazardous waste material. The toxic waste material surface impoundments are required in order to be manufactured with the following:
    - Double liner system
    - System for the leak detection
    - Leachate accumulation and the removal systems

In order to ensure that the proper installation and also the construction is done for the same, he regulations in this regard require that the unit needs to have and also follow the quality assurance program for its construction. The regulations are also known to outline the inspection, monitoring, response action and the ultimate requirements for the closure.

- **Boilers and the Industrial Furnaces** – The boilers just like the incinerators are also the enclosed devices which make use of the controlled type of the flame combustion in order to first recover and then export the energy in the steam form, heated gases or heated fluids. The industrial furnaces are also the enclosed devices, which are known as the integral component of the manufacturing process and then the usage of the thermal type treatment for the purpose of recovering any form of energy or the waste materials from the hazardous type of the waste. Some known examples of the industrial furnace could be the aggregate kilns, cement kilns and the halogen acid furnaces etc. where the last one is known to generate acid from the hazardous waste of the halogenated kind.
- **The waste piles** – The waste piles are known as the non-containerized piles of the non-liquid, solid or the semi-solid toxic waste materials which are used for the purpose of temporary storage or for its treatment. The waste files requires to have the following:
    - A liner system (double)
    - System for leak detection

  - The leachate accumulation and the removal systems (double)
  - The runoff, run on and the controls for the wind dispersal
  - Quality Assurance program for the construction

- **Tanks** – The tanks are the stationary units which are made from the non-earthen raw material and the facility is used for the storage or the treatment of the hazardous waste materials. The tanks used for the purpose are made from a variety of different types of materials which include the likes of plastic, steel, concrete and fiberglass etc.
- **Container** – The toxic waste material container is a kind of a portable facility which is used for storing, transporting, treating or may be handling the toxic waste material. The most commonly used waste container is actually the drum having capacity of 55 gallon. A few other commonly known containers for the hazardous waste material are the railroad cars, tanker trucks, test tube containers, bags and the bucket types of containers.
- **Injection Wells** – This type is primarily regulated under the known regulation of the Safe Drinking Water Act and the Underground Injection Control program of EPA.
- **Drip Pad** – A drip pad is actually a kind of an engineering type structure which consist of a free drainage base constructed from the non-earthen material type and that is designed for the conveying of the wood preservative chemical that drips out of the treated wood and the surface water runs over the associated collection system at the wood preserving plants.
- **Miscellaneous units** – The miscellaneous units for the management of the toxic type of waste material involve the units which do not fit well within the definition of the typical or the standard type of the waste management units/ sites or the facilities. Such miscellaneous or the other types include the repositories which are described below:
  - Underground caves

- Underground mines
- Units for thermal treatment which are other than the boilers, incinerators or the industrial furnaces.
- Salt bed formations
- Salt dome formations

As the units mentioned above are known to vary considerably, so for this reason, they are subject to their performance meeting the environmental standards and the mot prescribed standards which are technology based. The requirements of the standards is such that the facilities/ sites/ units should be located, designed, built, operated, run/ maintained and closed in such a way that there is no harm caused to the health of the human being or to the betterment of the environment. US EPA. 2017. *Hazardous Waste Management Facilities and Units | Hazardous Waste Permitting | US EPA*. [ONLINE] Available at: https://www.epa.gov/hwpermitting/hazardous-waste-management-facilities-and-units

THE IMPACT OF DIFFERENT TYPES OF THE WASTE DISPOSAL SITES

The impact of different types of the waste disposal sites could be considerable and deteriorating if the facility is not handled, maintained or managed well. On a broader view, the impact could be of 2 different types, namely, environmental and economical for the different types of the waste disposal sites. Each of them are explained below for further clarification:

1. **Environmental** – The following environmental effects could be seen with the different types of the waste disposal sites:

    a. **Soil Wastes** – Once any waste disposal site contains unattended or unprocessed toxic materials then those could get into the soil and may lead to impact the health of the plants when the contamination is consumed by the plants through their roots. When the human consumes those plants and the animals which in any way were known to be in contact of any form with the soil (pollutants) then there could be a seriously negative impact over the health of the human beings and for animals as well who consumes such plants for their needs.

b. **Pollution** – Any bad or low standard practices in the waste management process could lead to air and land pollution and that could also result in respiratory problems and multiple other types of the adverse health impacts, just as the contaminants into the lungs and then to several other parts of the body.

c. **Leachate** – The liquid that is generated as the water goes trickling through the contaminated places is known to be as leachate. It results in some really harmful mixture of chemicals which may cause the toxic substances to penetrate inside the ground water, surface water or the soil.

d. **Surface water contamination** – The waste that is leftover in the water bodies is known to change the composition of water in a negative manner. This in technical terms is known as, water pollution. This will eventually affect the entire ecosystem that lives/ exist in the water. It could also lead to harming animals and also the plants who consumes such water for their needs.

2. **Economical** – There are a couple of economic impacts as well for the waste disposal sites which are explained below:

    a. **The municipal wellbeing** – It is the basic need of every human being not only to live but also to visit places which are neat, clean, healthy and fresh for their health and overall wellbeing. Any city of the world that has poor sanitation standards or the waste disposal sites are underdeveloped or not standardized then this could lead to serious health implications of the human being also for the environment. Further, the smelly places with pollutions would not attract people of any sort, such as the tourist or the investors to visit that area for any purpose. The cities there are known to have poor living standards and the municipality wellbeing suffers big time.

b. **Recycling revenue** – Any city of the world which do not look to make any investment in recycling, reusability or the proper waste management processes then it misses out on a very big opportunity to generate handsome revenue from the recycling process of the waste material produced. The job opportunities that comes through the recycling activities are also denied or missed out. Apart from recycling, the opportunities for composting and several other business associated with recycling are also impacted financially which in the long run could be a big loss for the city, state or the country, at large. E School today. 2017. *Effects of poor waste disposal*. [ONLINE] Available at: http://www.eschooltoday.com/waste-recycling/effects-of-poor-waste-management.html

HOW THE MAINTENANCE OF THE WASTE DISPOSAL SITES COULD AFECT THE ECOSYSTEM?

The proper maintenance of the waste disposal sites could bring in a lot of positive impact on to the ecosystem. Besides that, the maintenance also offers lucrative revenue to be generated by designing and then operating such wosal sites. However, However, the effective maintenance brings a lot of benefits for the ecosystem in large and that eventually ensures that all the living beings are protective when it comes to the air, water and soil/ land that they interact with in their daily lives for numerous reasons. The following are some of the positive impacts that are gained with the effective maintenance or management of a waste disposal site:

- **It keeps the environment fresh and clean** – Human beings heavily depend on the consumption of water and the air that they breathe in. So, the freshness and purity of air, water and even soil is very important for the health of the living beings. The top most positive impact of the proper maintenance of the waste disposal sites is that it helps in keeping the environment neat, clean and fresh. The waste disposal facilities also enable the human beings to go disease-free as the resultawaste is disposed offof properly. The increase in the number of the waste disposal sites would further help the cause and that is also becoming a necessity with the growing needs in the waste disposal industry. Proper and effective measures are the key to success here. Just simply implementing a half-baked waste disposal facility would do no better to the ecosystem at all.

- **Conserves energy and saves the earth** – This impact on the ecosystem is mainly dependent on the recycling activity, which has become inevitable in the fast pace world of today. As the waste material recycling helps in the reduction of the cutting down of the trees, so that automatically helps the ecosystem to improve considerably. It is because the cutting of trees is mainly done for the production of furniture and paper and this could be considerably lowered, when recycling of the furniture products or the paper is done. On top of that, the recycling processes requires just a minimal energy for its complete processing as compared to making the same product afresh.
- **Reduces the environment pollution considerably** – What else could be done to the environment besides eliminating or at least reducing the pollution considerably. As explained in the above text that the proper management of the waste materials would not just help in eliminating the surrounding waste materials but it would also help in reducing the greenhouse gases intensity big time with, which include the likes of, carbon dioxide and methane etc., which is known to be emitted from all the hazardous waste material which is accumulated. The depth of the current incinerations and landfills would be curbed, which results in the cutting down of the toxic materials and other harmful factors which have the potential of impacting the environment in any way. Apart from this, the amount of the fossil fuels would also be significantly reduced by following this type which eventually would result in a very safe and greener environment to live/ breathe in. Sravani. 2017. *Advantages and Disadvantages of Waste Management - WiseStep*. [ONLINE] Available at: https://content.wisestep.com/advantages-disadvantages-waste-management/

WHAT MEASURES ARE TAKEN FOR THE WASTE DISPOSAL SITES TO WORK IN AN ECOFRIENDLY MANNER?

In general, the following measures are taken in the operations/ maintenance of the waste disposal sites to work in an ecofriendly manner:

- **Hours of operation and access control** – The access road to the vicinity of the waste disposal facility should be gated. The gate-house is ideally protected with the human force during the hours of operation. The gatekeeper at the gate should be interviewing the driver about the contents of the vehicle, fees/ charges and give him the right direction. This is the first step to ensuring environmental safety so everything is effectively managed for protecting the environment.

- **Spel arrangements forthe some specific wastes** – Some waste materials requires to have some special arrangement that done in advance;,Forfor example, the asbestos waste is required to be taken between the freeze-up and the break-up point and it is not to be accepted for disposal during winter season.

- **Sorting delivery** – The sorting of the waste materials based on its nature is done by putting them into segregated areas for the environmental concerns and the potential for recycling. The gate-keeper in this regard can instruct the driver to do the needful in the waste disposal facility.

- **Wastes segregated for recycling and/ or the special handling purposes** – There are multiple waste material types which are collected and stored separately and that is mainly done for their potential to be recycled or may be due to other maintenance/ management requirements. Some of the products that are stored separately for the purpose of recycling include the likes of beverage containers, furniture, building components, computers and peripherals and metals etc. Then there are some waste materials which due to their toxic nature and/ or the requirements for effective disposal needs to be treated and stored separately. The waste material in this regard include the wastes like, paint, latex, alkyd, freezers, refrigerators, tires, lead acid batteries and certain types of the building demolition material.

- **Handling waste material that is undergoing treatment** – There should be separated storage of the specific waste materials which are undergoing treatment. This include the soil that is contaminated with oil or also the snow which could also be contaminated with oil.
- **Scavenging regulation** – The public members who are holding a license (scavenging) could scavenge during the normal operation hours. No fee is required for this activity for those having the required license.
- **Burning** – Burning in the developed countries is prohibited for the waste material but this is also used sometimes to reduce the volume of the waste material.
- **Controlling the windblown debris** – The windblown plastic bags, paper and other similar types of light waste materials are trapped in the waste disposal facility to the best extent as possible by using the snow fencing.
- **Fire control** – Fires is occasionally initiated inadvertently or it could be as a result of vandalism. This is dealt by the fire department and arrangement in advance should be there for any emergency situation.
- **The contingency planning and the environmental surveillance** – The monitoring of the drainage is done for ensuring the quality of the waste disposal site. A flow minimization strategy should be in place where the surveillance sampling done indicates some level of contamination passing on. http://registry.mvlwb.ca/Documents/G17L3-001/G17L3-001%20-%20Solid%20Waste%20Disposal%20Facility%20OM%20(2012)%20-%20Feb06_17.pdf

Waste management has always been a big talk when it comes to the quality or the living standard of a city. It is not that every city for its better environment should have the in-city waste disposal sites. The handling of solid and/ or the semi-solid waste material needs some solid efforts for the effective disposal or overall handling of the waste materials. Rather, the need is to have a proper mechanism of storing, treating and disposing the hazardous waste materials which could be anywhere in or may be out of the city where a waste disposal facility is located. Several waste disposal facilities such as landfills, incineration, containment building and the eland treatment units etc. have different structures and the operational processes for the purpose of accumulating, processing and/ or the proper disposal of the waste material. The suitability for a particular type of the waste material disposal site depends on the nature of the waste material and also the ease of access of the facility for a particular purpose. The most effective handling of the waste material of any kind requires to make the recycling processes as practical as possible. If recycling is done to the maximum for the amount of waste material produced then that would surely be beneficial for the environment and the health of the human beings.

7 SUSTAINABLE WASTE MANAGEMENT

Sustainability is what that is more focused and prioritized in the various forms and almost each and every field where there is a need to use resources from the natural environment, there is always a focus on the various elements and procedures that make sure that there will be processes and steps that will assure the sustainability of the environment and the natural resources that are used in completing the process.

Same is the case with the waste management procedures. When it comes to finding out the various ways that are used in managing and treating the various kinds of hazardous wastes, there is always an effort to make things possible so that there are no issues in managing the waste materials and the byproducts as well. It is also made sure that when the products are wasted, they should be treated to give useful products that in turn will have given rise to the production of the new materials, as well as the environment that stays safe for all.

The sustainable management of the various wastes and materials that are hazardous and not biodegradable involve the various careful steps that make sure that the processes and the byproducts as well as the end product would be biodegradable and there will be no extra products formed that might affect the resources or the environment as well.

The main elements of introducing the hazardous wastes in sustainable action

There is a great need to find the various materials as well as the processes that make sure the sustainability of the various procedures possible for most of the times. The most important things that might be involved could be:

- Advanced technologies to reduce wastes

It is surely said that when there is a need to cater to the increasing needs of hazardous wastes, then there should be a clear vision of considering all the hazardous waste by using the latest technologies. As a fact,the latest procedures tend to provide the best possible solution for managing the hazardous wastes through proper processing methods.

- Increasing the recycled products and reusing the products later on

It has also been noted that reusing the various products that can be reformed and reused in various ways also contribute in the production of useful products without increasing the overall cost and help in the sustainability of the resources and the various components as well.

- Creation of even safer processing and treatment and disposal options

There is also a great need of introducing the various substances and processes that may help in producing the various procedures that not only help in the recycling of the products but it may also help in managing the wastes by applying safer, better and more sophisticated ways to make sure the overall process swill not harm the ecosystem at all.

- Emergence of the resources to manage the renewable energy

Another element that takes part in the development of the various processes regarding the renewable energy. Because of the fat the proper deposition of the wastes also encourage the production of renewable energy and may also give rise to the options that clearly support the usage of renewable energy so that there is lowered risks of running out of our natural resources that might be used during the process of waste management.

THINKING GREEN FOR MANAGING HAZARDOUS WASTES IN THE VARIOUS DEVELOPING COUNTRIES

- Always make sure to use the resources that are clean, renewable and are easy to access without creating any kind of mess and possible hazards.
- To reduce wastes, reuse the possible recyclable items and recycling is possible, there should be no reason to avoid it.
- Cleaner and bettertechnologies with lesser complication should be preferred instead of old worn out ways to deal with the hazardous wastes.

There should always be an emphasis to manage things properly and to give the best sustainable system so that there are no chances of getting the ecosystem in jeopardy and still makes enough progress in getting rid of the various kinds of wastes.

SUSTAINABLE WASTE MANAGEMENT SYSTEMS GUIDELINES AND PRINCIPLES

SWMS Guidelines and Principles

Achievement	Lead by example with sustainable policies and practices
Dedication	Seek integrated waste management solutions including diversion
Collaboration/Transparency	Facilitate cooperation and collaboration, including encouraging partnerships
Service to our Community	Engage the community to ensure the SWMS meets their needs and objectives
Sustainability/Integrity	Recognize waste is a resource and find solutions that support reduction, reuse, recycling and energy recovery before disposal

Continuous Improvement	Invest in infrastructure with the flexibility to accommodate growth, changing policies and encourage innovation

Source: http://www.barrie.ca/Living/GarbageAndRecycling/Pages/SustainableWasteManagementStrategy.aspx

ADVANTAGES OF SUSTAINABLE WASTE MANAGEMENT

Due to the facts, people, are now moving towards the lifestyle that is the most valuable, easy going and is harmless to all of the ecosystems and living being around them. They always need a solution to their problems in a way that they would benefitted a lot and others as well or if not there should be no harm done to others including the environment as well.

There could be many different kinds of advantages that could be related to the various benefits that are the result of sustainable hazardous waste management disposal.

The following are some of the important key points:

The less use of energy resources

When we are working on a sustainable procedure that takes care about all the expenses and all the various kinds of processes to make sure that there would be no extra burden in making things simpler. Rather the processes would be formulated in a way that no one would be harmed and all things could be sued for the betterment of each other and the environment as well. So, in that case when it comes to the energy resources and you are using the biodegradable substances you can surely lower the cost of maintenance while still they manage each and every process easily.

Least cost of getting new products which might have to be imported

When there is a complete setup of having the used products get used properly again, there would definitely be a least cost involved in the production and processing of the new products that in turn reduces the overall cost and the need for obtaining new raw materials where all the materials are reused immediately.

Lowered cost of producing new materials

When the materials are obtained in the form of the recycles materials, there would be lesser cost of obtaining new products hence would be better to use instead of purchasing all the new products every time that they need.

ESSENTIAL METHODS OF SUSTAINABLE WASTE MANAGEMENT

The basic and the most essential components of the various kinds of sustainable waste management procedures involve the various stages through which the disposed products are either reused in a different way or the products that are available must be given a new shape and form to provide the best possible products or the byproducts that take the useful form and give rise to more better components that are reusable and are less hazardous to the environment and may not give rise to serious issues in the ecosystem.

It is better not to convert things using the various expensive products as well as the one which in turn give rise to further issues in the environment. Such products are not preferred because the concept of sustainability requires, high quality but easy and cheap methods and components to be used that are never hazardous to the environment as well as to the community that has created and trying to reuse it for once.

It is therefore kept in front that when the products are reused, they are given the form that is most sustainable and are not dangerous to anyone who is either using them or they have converted and given a new form for the sake f waste disposal, recycling and reuse of the products.

The basic methods that are followed in the various waste management techniques are:

Reuse is the first and the foremost process that is put to work for the purpose of reusing and maintaining a particular kind of waste that can be used either in the same condition or it can be treated and further established to be reused in the various manners.

In the same way, if the products that have been disposed of had to be reused, they might have to have some sort of maintenance and given a certain form of shape or structure to give it a more proper forms that it can be reused in the same way as it was done before.

Recycle

Further if the products cannot be used as it is and they have to be treated properly before they can be used and reformed for various purposes or they might have to break them down to give them a new composition that the process we use is called the recycling.

During the recycling process, either the whole product is broke down into its smaller components or they might have to go through a prolonged process and time period to give it a new structure that is more environmentally friendly and useful for the users and may not harm the environment in any way.

Energy recovering process

In case if the recycled materials has the capability to provide certain levels or certain forms of energy then it should be utilized or converted into the various forms of renewable energy so that there would be least loss of useful compounds as well as least production of hazardous wastes that have to be disposed completely into the landfill and other such available waste disposal sites.

It is therefore important that when there is a need to manage a sustainable program of waste management where the various forms of hazardous waste management techniques are used then there is a high need of decomposing the compounds in a non-hazardous manner so that they can be used as useful products in various ways and may not harm the ecosystem and the users and can be sustained as an alternate form of energy that is sustainable and renewable for various purposes and for a long time until and unless it has to be disposed of in a careful manner without destroying the environment in any way.

Landfill and complete disposal

After the usage of these kinds of waste products and conversion of the products from wastes to the renewable energy or products, when the leftovers are to be disposed of, they need to be treated properly through various procedures to make sure they will be made biodegradable and may not cause any further issues no matter where they have been disposed and they could be disintegrated properly in a natural manner.

ELEMENTS OF
A WASTE MANAGEMENT SYSTEM

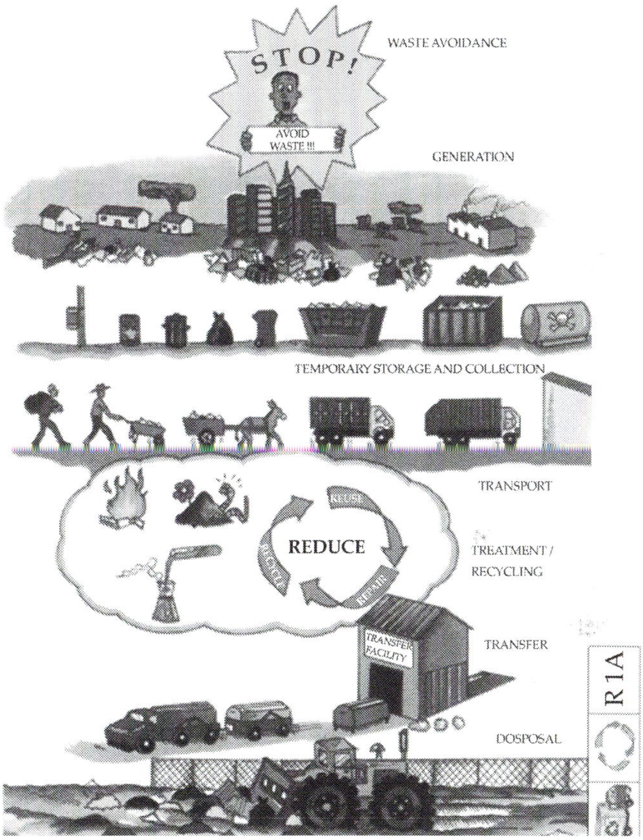

Figure 2: https://discardstudies.com/2013/11/21/cfp-sustainable-waste-management-systems-aag/

The sustainable waste management involves the environmental, economical as well as the community level affects when it is being processed to give the best possible solution for the sake of the betterment of the society and at the community level as well.

Due to the fact when the environment has to be maintained through various tactics used in the sustainable waste management system, it is being processed to maintain the main components and the balance within the environment so that the equilibrium in the ecosystem is not changed or altered ina negative manner. This ensures that the components in the environment as sustained by using or disposing the waste materials properly. Though for the sake of having all things support a sustainable system there might be a need of having a proper plant or the system that guarantees the production of the useful wastes andthe usage of the wastes in the useful manner.

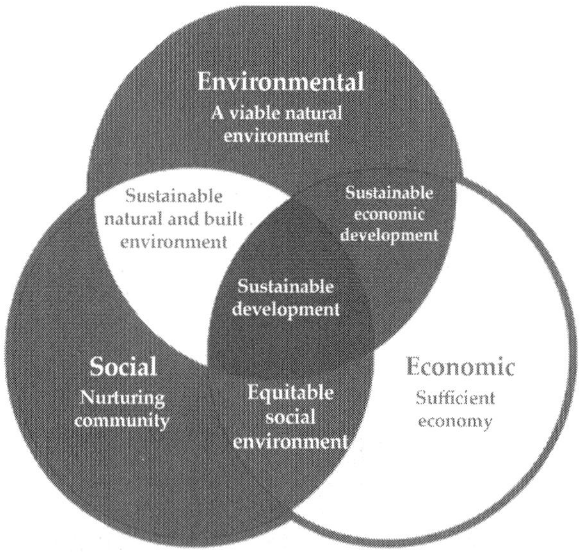

Figure 3: http://asiwaste.com/solutions/sustainable-waste-management/

In another ways, when we say that we have to manage the sustainable waste management program for the sake of managing the best practices in our society and at an economic level, we must keep that mind that a sustainable socioeconomic development can easily be maintained with the help of using the waste materials in productive manner and handling all the various related task in a careful way so that the environment does not get affected by the various components that are being processed or used in various ways.

It is therefore said that when there are many techniques that are used for the maintenance of sufficient economic resources, proper and sustainable waste disposal can play an important role by decreasing the factors that hinder the way to succeed in different financial goals as well as through various procedures that may give rise to acceptable procedures to support the economic condition of the country.

It is therefore very important and a crucial step forward to support the various processes in the developing countries that is the waste disposal is done via using the sustainable waste disposal measures, the process of economic progress is supported equally.

Further the nurturing community develops when there is a sustainable waste management system that supports the energy renewal processing and it give rise to a sustainable community with least hazards and high end productivity with the usage of best practices for every kind of waste disposal and recycling procedures.

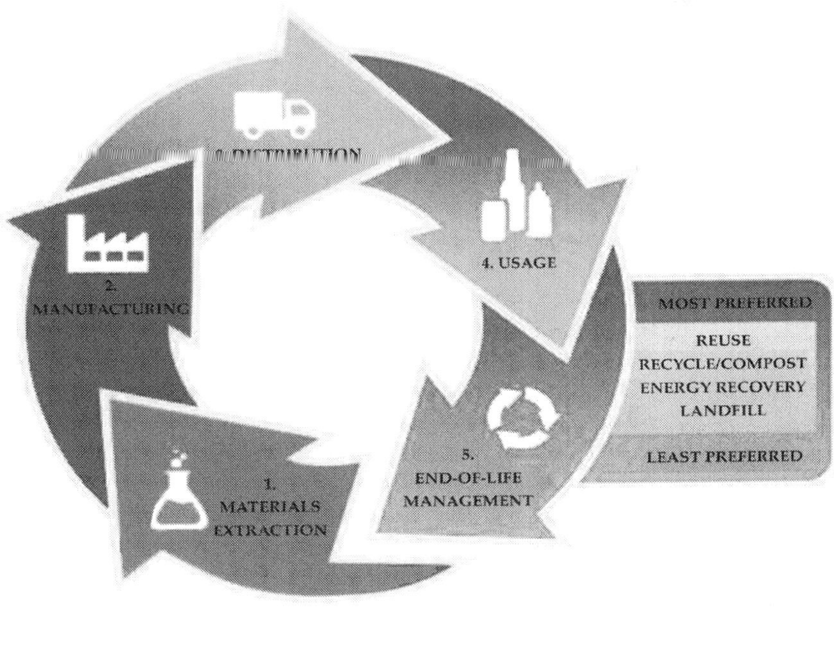

Figure 4: https://www.epa.gov/smm/sustainable-materials-management-basics

THE INTEGRATED SUSTAINABLE HAZARDOUS WASTE MANAGEMENT

Figure 5: http://wasteportal.net/en/waste-aspects/integrated-sustainable-waste-management-iswm

The integrated sustainable waste management system has been one of the preferred mode of waste management in the societies where there is a need to give rise to a sustainable system that not only supports the economic conditions but also facilitates the various procedures that make sure that there would be no harm done to the society and also the environment.

In most cases, the basic stakeholders of an integrated system in a society for the sake of management of the waste materials are:

- The local government and authorities
- The NGOs
- The various service users
- The private sector workers
- Donors and supportive agencies

The basic reasons why these agencies and the various government and non-government sectors work to provide an integrated system of the sustainable waste management system is to support a sustainable society that is sufficient enough to maintain the eco-balance and to help the society in improving the various waste management conditions that plays an important role in developing better societies.

The role of these services is to make sure that the system gets sufficient support from the finance sector and can also improve the overall financial situation to support further processes in a productive manner.

It is therefore important that the sustainable integrated management of the waste products in any society has to be managed and looked after in a proper way by knowing the various management procedures that affect the processing and management of the waste directly or indirectly.

Meaning of Integrated Waste Management

This section shows a concise outline of the historical use of the term 'integrated' related to solid waste administration and navigates its specific meanings. The term had turned out to be standard by the mid-2000s, as suggested by its wide adoption by the research group and reflected in names of recently establishing waste-related scholarly research centers such as 3R: Residual Resources Research, a PhD research school on integrated resource and waste administration of DTU, Denmark, the Integrated Waste Management Center of Cranfield University, UK, Integrated Waste Systems of The Open University, UK, the CSIR Center for Integrated Waste Management, South Africa and the Center for Integrated Waste Management of the University at Buffalo in the USA. To the best of the authors' information, in the re-assessed literature the term 'integrated' was first associated with solid waste administration during the 1970s (Tobin & Myers, 1974) and there has been an exponential increase in use of the term since at that point. A search on the Scopus research reference database did in March 2012 uncovered just 10 papers published with every one of the terms 'integrated' 'waste' 'administration' in the title by 1980, 31 by 1990 and 203 by 2010.

Nonetheless, every one of these uses a systems hypothesis approach, separating out identifiable discrete entities ('items', 'elements' or 'units') to describe relationships among them. Most of these uses reflect two of the lexicon meanings of 'integrated':

- combined or composite, made up of parts that function admirably together or
- Combining separate things, bringing together processes that are typically separate.

Be that as it may, some uses do also mirror the third word reference meaning

- Open to all individuals, as in integration by, for instance, race, ethnicity, religion, gender or social class, which somehow relates to the social/human capital of sustainable improvement.

Most of the terms used for 'integrated waste administration' and 'integrated solid waste administration' are to a great extent specialized, focusing on the best way to integrate various specialized elements into a more total or potentially provincial system, and sometimes on using PC helped tools to assist with that integration. Most of the references are also identified with developed countries. A vital variety interprets integration in the sense of addressing the greater part of the levels of the waste hierarchy, from counteractive action through recycling to disposal.

The thought of sustainable advancement came to worldwide prominence by method for the report Our Common Future, published by the World Commission on Environment and Development (WCED) in 1987, however it consolidated on and interpreted precursor versions (Mebratu, 1998). Motivation 21 – the activity anticipate sustainable improvement concurred at the United Nations Conference on Environment and Development in Rio de Janeiro in 1992 – has been influential in solid waste administration. Diaz et al. (Diaz, Savage, & Golueke, 1996) contended for a holistic and systems way to deal with 'integrated and sustainable solid waste administration (SWM)', addressing every one of the three aspects of triple main concern interpretation of sustainability (natural, social and financial) with a specific emphasis on spatially combining solid waste administration, wastewater treatment, energy production and food production facilities.

Origins of Integrated Sustainable Waste Managament

Interestingly, the characteristic of integrated waste administration, being interested in every one of the actors or stakeholders according to the third word reference definition seems to have been explicitly taken up essentially with regards to developing countries. In the mid 1990s, numerous international agencies and nongovernmental organizations (NGOs) dynamic in developing countries ended up plainly disenchanted with the disappointment of the traditional, exclusively specialized, approach (frequently called the 'technical fix' (Wilson, 2007). A collaborative program on metropolitan solid waste administration in low-income countries was set up by UNDP, UN-Habitat and the World Bank, with funding from the Swiss Agency for Development and Cooperation. A workshop gathered in Ittingen, Switzerland, in 1995 resulted in a theoretical structure for integrated metropolitan SWM in low-income countries (Schu¨ beler, Wehrle, & Christen, 1996). This novel approach essentially based upon the holistic, comprehensive, all-inclusive idea of sustainability on various levels. It distinguishes between three principal dimensions, each indicated by a question:

- What? – the scope, which includes the physical components of a waste system however goes significantly facilitate by identifying various other planning and administration issues that should be addressed, such as strategic planning, open investment, financial administration et cetera.
- Who? – focuses on the stakeholders or actors, according to the third word reference definition.
- How? – How strategic objectives and issues should be addressed introduces interestingly a series of strategic aspects: political, institutional, social, financial, monetary and specialized.

Some of these are also recognized as forming the more extensive nearby setting – political, financial, socio-social and ecological – that needs to be considered. Some of the early thinking underpinning the reasonable system can be credited to Arnold van de Klundert of the Dutch institute-sort NGO WASTE. The Dutch government supported WASTE to attempt a six-year Urban Waste Expertise Program (UWEP) from 1995, which fabricated further on the applied system to create what was surprisingly named integrated sustainable waste administration (ISWM) as both a systematic apparatus and an advancement structure (Van de Klundert & Anschu¨ tz, 2001).

As shown in Figure 6, this maintains the three dimensions, yet simplifies the ‹What?› into just the physical components (elements) of the system.

Through the 2000s, the idea of ISWM was additionally refined and is bit by bit becoming the standard in discussion of solid waste administration in developing countries. A second phase of UWEP resulted in a guideline on using the ISWM assessment philosophy. An EU-supported program on ISWM in Asia (www.issowama.net) used ISWM to build up a task specific way to deal with assess average success and disappointment factors. The original international collaborative program advanced into a collaborative working gathering (CWG) on solid waste administration in low-and center income countries – various international workshops have additionally elaborated on ISWM (www.cwgnet.net). The CWG shaped the foundation of the 35-strong international group who arranged Solid Waste Management in the World›s Cities for UN-Habitat (Scheinberg, Wilson, & Rodic, 2010b); this adjusted the three-dimensional ISWM structure, for the purposes of a systematic comparison of the cities, into what can be described as two overlapping triangles (Figure 7).

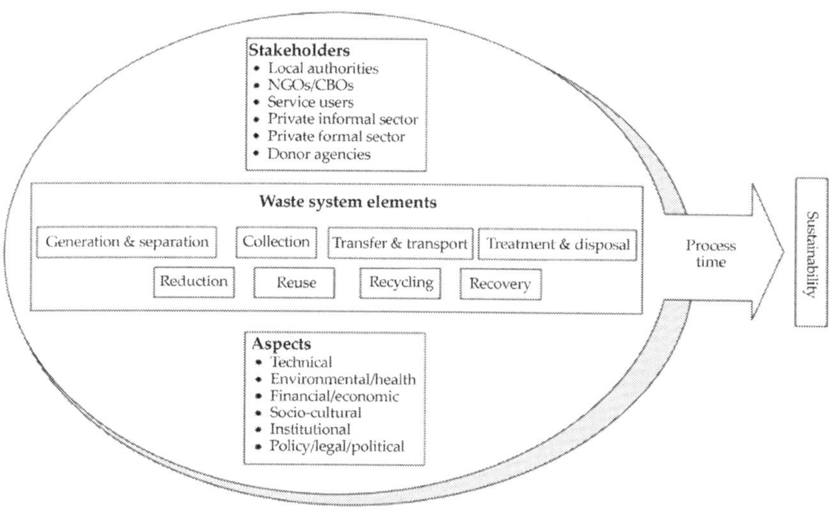

Figure 6: Waste System Elements

Source: (Van de Klundert & Anschu¨tz, 2001))

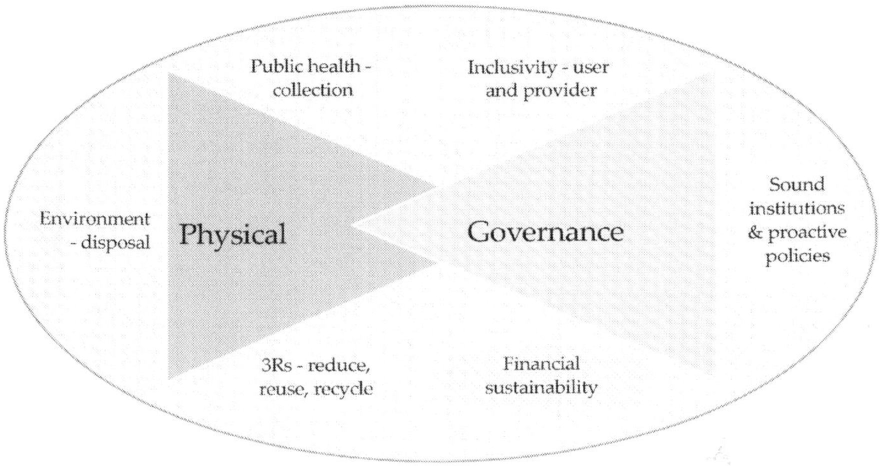

Figure 7: Two Triangles Representation

Source: (Scheinberg, Wilson, & Rodic, 2010b)

REALITIES AND CHALLENGES

Public Health

The careful extraction and subsequent administration of solid waste sits close by the administration of human excreta (sanitation) as two of the most essential urban ecological administrations. The essential driver is general health; cholera epidemics in the nineteenth century prompted genuinely far reaching, municipal solid waste gathering administrations in significant European and US urban areas by 1900. Uncollected solid waste is as yet a genuine general health issue in developing nations. Health information indicates essential higher rates of diarrhea and acute respiratory diseases for kids living in families where solid waste is dumped or burned in the region, as contrasted to family units in similar urban communities that get a general waste gathering service. Uncollected waste likewise pieces drains and causes flooding and subsequent spread of water-borne ailments: this was the reason for a noteworthy flood in Surat in India in 1994, which brought about a flare-up of a malady influencing 1000 individuals and causing the death of 56 (Gupta, 2010).

Information on waste composition indicates high plastics substance of solid waste irrespective of income level of a nation (normal crosswise over 20 reference urban communities: 10% on a weight-for-weight premise). The yearly floods in low-lying parts of Kampala, Uganda are faulted to some degree for plastic packs, privately known as buveera, hindering the drains. In West Africa, floods have been faulted for the uncollected little plastic pouches in which drinkable water is sold. After the overwhelming floods in 1998, when plastic packs obstructed drains and postponed water levels falling, Bangladesh was the main nation on the planet to boycott plastic sacks in 2002. So also, in light of the disastrous outcomes of the rainstorm flooding in Mumbai and the whole territory of Maharashtra, India in August 2005, including more than 900 passings and billions of rupees (a huge number of US dollars) in material harm, the province of Maharashtra banned the produce, deal and utilization of plastic sacks as they were found to have obstructed the drains and along these lines exasperated the circumstance. The most slender plastic packs have likewise been banned in a few East African nations (BBC, 2007).

Waste accumulation is a merit decent – a great (benefit) considered so critical that the law requires that it is accommodated for the benefit of the whole society, paying little respect to the enthusiasm of the market to supply it or the clients' capacity (or eagerness) to pay for it. The key marker here is gathering coverage or the level of the populace that approaches waste accumulation services – in principle, this ought to be 100%. Starting in July 2012, The World Bank site was all the while expressing that it is 'normal that 30– 60% of all the urban solid waste in developing nations is uncollected and under half of the populace is served' (World Bank , 2012) – which was the situation in the 1990s.

Throughout the most recent decade, urban areas have attempted impressive endeavors to increase benefit coverage: half of the reference urban communities, including everything except two of the urban areas with a human development index (HDI) of at least 0.75 (or a gross national income (GNI)/top above $1600, report coverage rates of 99–100%. The poorest performing of the center income urban communities have gathering coverage in the scope of 70– 90% populace served, while every one of the six of the low-income reference urban areas demonstrate accumulation coverage in the range 45– 60%. Our decision on late advance is reinforced by a current World Bank report that gives normal accumulation coverage of 86% in upper-center, 68% in lower-center and 41% in low-income nations – these figures are to some degree lower than those recommended by information gathered by the present creators; the figures originate from a significantly bigger example measure, however are to some degree more established, with a median date of 2001.

In any case, every single such figure hides the gap between the 'haves' and 'have-nots'. Inside numerous urban areas, the focal business region and affluent neighborhoods have close to 100% coverage, while low-income and illicit settlements frequently have none. This obvious gap in the execution of the minimum developed urban communities implies that enhancing gathering must even now be their first priority, together with water and sanitation, if general health is to be protected.The barriers and requirements to expanding service coverage are regularly predominantly identified with the 'delicate' administration issues, as talked about later. Tailoring administrations to nearby conditions is especially vital. In casual settlements with poor access, essential waste accumulation utilizing hand, creature or bicycle carts will probably be proper – modernization does not need to mean motorization. In Kunming, China, tricycle carts convey waste to more than 120 little transfer stations where the waste is then transferred to modern vehicles for optional transport to transfer destinations; comparable little transfer stations are getting to be noticeably basic in Vietnam and Egypt (Coffey & Coad, 2010).

A noteworthy distinction between nations is in waste composition: organic substance in developed nations midpoints to a 28% while that in developing nations is around 67% (Wilson, Rodic, Scheinberg, & et, 2012). This influences developing nations to waste essentially wetter and denser, so compaction is for the most part not required to accomplish the greatest vehicle stacks that the nearby streets can convey. The utilization of imported vehicles, for which spare parts are regularly inaccessible, and bad infrastructure for maintenance are wellsprings of framework failure – it is normal for half or a greater extent of a city's accumulation fleet to be out of service at any one time.

Environment Protection

At the point when the period of solid waste modernization started in the 1960s, the significant driver was recently rising ecological concerns. The developed nations continued in a progression of steps (Wilson, 2007), first eliminating or redesigning public dumps to 'controlled disposal' and after that bit by bit expanding the models of leachate and gas control. Advance was moderately slow, so that by 1990, uncontrolled disposal was as yet huge in five of the then 12 EU part states, with levels more than half by weight in three. In July 2012, the World Bank site announced that 'in most developing nations, public dumping with public burning is the standard' (World Bank , 2012) – this was the situation in the 1990s. In any case, late information again recommend that the present reality constitutes a critical change.

The driver to enhance waste disposal and additional develop engineered landfills has chiefly been ecological security, yet frequently it has taken a fiasco to draw the consideration of both the general population and government officials. In the USA and Western Europe in the 1980s, the impulse was given by broadly advanced occurrences of uncontrolled unsafe waste sites, for example, Love Canal. In various developing nations, there have been real mishaps including landslides on shaky waste slopes at uncontrolled solid waste dump sites murdering tens or many individuals, for example, in Istanbul, Turkey in 1993, Manila, Philippines in 2000 and Bandung, Indonesia in 2005. As of now, because of big protests from natives and the sites ending up full, megacities, for example, Delhi and Mumbai are shutting down their current dump sites (Rodic & Gupta, 2012).

Waste disposal remains a noteworthy issue for some low-income nations. The barriers and constraints to enhancing disposal are frequently administration issues. Both high capital speculation and high working expenses are key barriers. Capital expenses are frequently tended to by seeking global development aid subsidizing (e.g. Lusaka, Zambia – Danish subsidizing by DANIDA; Dhaka, Bangladesh – Japanese financing by JICA). Be that as it may, most aid offices won't add to continuous working costs, so a city's capacity to oversee operations and pay these expenses is critical. A few urban areas in low-income nations (e.g. Bamako, Mali) have delayed the development of an engineered landfill partly because of the uncertainty issue of funding for transport and operation. Different cases incorporate the Jam Chakro landfill in Karachi (Rouse, 2006), where a state of the art landfill site worked with aid funding has immediately returned to an uncontrolled dump due in any event in part to an absence of funds for appropriate operation. In this specific situation, carbon-development mechanism for methane recuperation from landfills has been critical, as cash is just paid out when proof of legitimate operation has been given. Kunming in China and Belo Horizonte in Brazil are among the urban areas that have done this, yet it could be all the more generally utilized.

Safe disposal by method of thermal preparing combined with vitality recuperation and the various varieties of mechanical biological treatment plants assume a noteworthy part in modern advanced solid waste administration in developed nations. Much like in Europe in the 1960s, 1980s when modern thermal preparing advancements were developed and connected to treat developing measures of waste, substantial Chinese and Indian urban communities are thinking about or settling on thermal handling of their municipal waste. Such advances, nonetheless, are yet to establish their positive reputation under the common conditions in developing nations, where municipal solid waste is high in natural substance and subsequently moisture, along these lines perhaps rendering waste unacceptable for thermal processing without pre-treatment or the utilization of fuel, and where affordability and institutional capacities with respect to satisfactory operation, maintenance and emissions observing are real issues still.

3 R's – Reduce, Reuse, Recycle

Before the industrial revolution, most urban communities had a couple of material assets, cash was rare and family units had a larger number of necessities than they could meet. The predominant driver for what we now call solid waste administration was the market value of materials and items: wastage was limited, items were recycled and reused, materials were reused and the organic issue came back to the soil. Extensive private, entrepreneurial (or 'informal') recycling frameworks flourished. Developing formal municipal waste accumulation frameworks in the late nineteenth century displaced these informal recycling frameworks, which, in mix with expanding accessibility of merchandise from shabby large scale manufacturing, added to a decrease in recycling rehearses. Except for the two world wars, recycling rates from postconsumer municipal solid waste step by step declined to low single-figure rates by 1980. Since the 1980s, high-income nations have been rediscovering the value of recycling (which in this setting incorporates fertilizing the soil) as an essential part of their SWM frameworks, and have put funds in both physical infrastructure and correspondence strategies to increase recycling. By 1990, the Netherlands and Germany had remade recycling rates to 15– 20%wt, with a normal of 8%wt over the then 12 EU part expresses; the normal recycling rate over the 27 EU individuals in 2009 was 42%wt (Eurostat, 2011). The driver is not fundamentally the commodity value of the recuperated materials, which was the main inspiration of the prior, informal or private sector, frameworks. Or maybe, recycling markets offer an aggressive 'sink' as a contrasting option to progressively costly landfilling, thermal preparing or other treatment choices.

Many developing and transitional nation urban communities still have dynamic informal sector and miniaturized scale undertaking recycling, reuse and repair frameworks, driven completely by the market value of the materials and (disposed of) items. Additionally, by taking care of such huge amounts of waste that would some way or another be gathered and discarded by the city, the informal recycling sector has been appeared to spare the city maybe 20% or significantly a greater amount of its waste administration budget. Essentially, the informal recycling sector – in a lot of situations the city poor – is subsidizing whatever is left of the city. There is a noteworthy chance to expand on these current recycling frameworks

- to increase additionally existing recycling rates
- to ensure and develop individuals' livelihoods
- to address the genuine and seen issues of such actions (work related and general health and well-being, child labour, uncontrolled contamination, untaxed exercises, wrongdoing and political plot)
- to decrease still further the expenses to the city of overseeing lingering wastes.

These difficulties of integrating the informal sector with the formal can be tended to, yet just if a systematic approach is followed and mutual troubles are transparently acknowledged. Oddly enough, there is some proof that recycling rates are lowering in a portion of the more developed upper-center income nations, maybe mirroring the historical backdrop of the developed world where the early formalization of solid waste administration as a municipal administration displaced prior informal recycling frameworks. The execution of a few nations in the developed world on the majority of the benchmark indicators talked about so far – gathering coverage, controlled disposal, recycling rates – has been exemplary, so one may presume that they have 'settled' the solid waste administration issue. Additionally, the issue has intensified over late decades – normal family unit waste era in the EU was 329 kg for each capita every year in 1990 and for 12 nations, municipal waste era in 2009 arrived at the midpoint of 591 kg (Eurostat, 2011). Allowing for conceivable changes in definition, the increase is extensive – regardless of late abatements in a few nations, which may reflect endeavors to decouple waste era from monetary development, and in addition the impacts of the financial retreat.

The needs of good waste administration, at the highest point of the waste hierarchy, are communicated by the '3Rs' – reduce, reuse, and recycle. Waste prevention, or decreasing the amounts of waste being generated, may in the long run turn into the new concentration of modernization in developed nations. On the off chance that development proceeds on late patterns, at that point a mix of populace development, rural– urban migration and increases in per capita waste era are conjecture to twofold municipal solid waste amounts in center and low-income nations by 2025, so they too should be in any way guiding consideration regarding bringing waste development rates under control. Reuse – specifically or following repair, restoration or re-make – broadens the helpful existence of items. Recycling can be conveniently part between 'dry' recyclables, which can be separated, prepared and come back to modern value chains, and natural wastes. The last incorporates plant and creature wastes from open green spaces, kitchens, gardens and urban agriculture, together with securely overseen and treated human excreta. These are wellsprings of organic matter and key supplements for the agricultural value chain, and their appropriate use by method for use as suitably prepared creature food, treating the soil or anaerobic processing is critical to soil essentialness and food security (Stuart, 2009).

8 SOLID WASTE MANAGEMENT IN DEVELOPING COUNTRIES

INTRODUCTION

As urbanization keeps on occurring, the management of solid waste is turning into a noteworthy public health and ecologic worry in urban territories of many developing nations. The problem is no joke, particularly in the capital urban communities, which are frequently doors to the nations for remote diplomats, representatives, and voyagers. Poor visual appearance of these urban areas will affect authority and tourist visits and foreign investment.

Knowing its significance, various developing nations have asked for joint effort of external help organizations, both bilateral and multilateral, in improving solid waste management in their urban areas over the most recent 20 years or somewhere in the vicinity. Albeit a few ventures prevailed with regards to giving steady positive effects on the management of solid waste in the recipient nations and urban communities, many neglected to proceed with exercises after the external help organizations ceased their help. This unsustainability of collaborative activities is because of different technical, financial, institutional, monetary, and social constraints looked by both the recipient nations/urban areas and external help offices. Such constraints are different from nation to nation and from city to city, as developing nations and urban areas inside them differentiate in solid waste management issues and they and external help organizations have unique, and frequently restricted, assets accessible to determine the issues. In this manner, keeping in mind the end goal to ensure the sustainability of collaborative ventures, the different constraints of both developing nations and external help offices ought to be painstakingly inspected and an approach to be developed to expel such constraints inside the setting of the collaborative undertakings. This paper delineates regular such constraints and proposes conceivable methods for evacuating these constraints.

SOLID WASTE CHALLENGES IN DEVELOPING COUNTRIES

Waste Problems reflecting the state of Society

Waste issues are said to be the reflection of a society (Sakurai, 2000). The condition of a society is firmly identified with its monetary, historical, social, ecological and other angles. These perspectives vary contingent upon the nation, city or group, as do waste issues. Understanding the condition of a society along these lines gives a course to a comprehension of the waste issues in that particular society. On the other hand, it is conceivable to comprehend the condition of a society by taking a gander at its waste issues. For instance, shell mounds piles of shells or kitchen middens left by shellfish eaters in prehistoric circumstances give pieces of information to what that prehistoric society resembled and how individuals in that society lived.

Swinging to the historical backdrop of waste administration in Japan, waste issues have for sure mirrored the changing territory of Japanese society. In and after the high development time of the mid-twentieth century, industry produced items and buyers grasped wasteful utilization examples and ways of life in light of the paradigm of large scale manufacturing and massive waste. Japanese society took this consistently expanding volume of waste allowed regularly even with hard to-treat waste containing different chemicals. The nation's endeavors to increase the quantity of treatment/disposal sites neglected to find up with the expanding volume of waste, allowing unlawful disposal to end up plainly across the board. Strict measures to control water and air contamination in the midst of developing worry about the earth brought about increases in sludge, soot, dust and other solid modern wastes, putting an extra strain on the nation's waste disposal limit. In spite of nearby endeavors, for example, the arranged accumulation of combustible, noncombustible and recyclable wastes, Japan relied upon the development of more incineration plants and other disposal sites to adapt to the development in solid waste. Japan's emphasis on incineration as a methods for waste treatment prompted more incinerated waste than in some other nation. With respect to modern waste, the nation fixed directions and punishments for disposal temporary workers (Tanaka, 1996).

Nonetheless, the main drivers of solid waste issues are really identified with the profligate era of solid waste and the large scale manufacturing and mass utilization of items without thought for their last disposal. There can be no answer for the issues of solid waste until the whole flow of the procedures of production, utilization and disposal are considered as an integrated framework. Also, the proposed development of new incineration facilities or landfill sites have turned out to be greatly disputable with restriction in light of the NIMBY (not in-my-backyard) principle and the developing impediments on the aggregate sum of accessible limit at current landfill sites. These were the extra factors that provoked the look for a more crucial answer for solid waste issues. Along these lines, before the end of the twentieth century, Japan had at long last started to manage the solid waste issue including all fragments of the society. These endeavors evolved into the idea of a "recycling-based society," because of which the Basic Law for Establishing a Recycling-based Society was authorized, followed by real modifications of the Waste Management and Public Cleansing Law and the Law for Promotion of the Effective Utilization of Resources and also the enactment of a few item particular recycling laws. This to some degree acerbic history of waste administration in Japan can be viewed as an impression of the progressions that occurred in the condition of the nation's society from the post-war time of high monetary development, to a world financial power, at that point to the fall of the "bubble economy" and the ensuing stagnation, lastly to the present procedure of adjustment.

Much the same as a stream running from the mountains upstream down to the ocean, when human produce and consumption material (upstream process), society definitely generates increasingly solid waste (downstream procedures). At the end of the day, there is a metabolism happening in materials amongst nature and human society. In any society, along these lines, solid waste is discarded in somehow and this administration framework has evolved and has been enhanced after some time. Watchful perception will uncover that such historically evolved waste administration frameworks exist in any developing nation. In any case, in parallel with the drastic extension of upstream exercises, for example, that accomplished by Japan in the last part of the twentieth century, developing nations in particular have additionally observed their customary waste administration frameworks overwhelmed as the objective of "development" was sought after too rapidly, there was insufficient care in arranging and a consideration regarding issues related with urbanization and populace fixation which developed at an accelerated pace. Clearly, a disparity had emerged in different nations between the condition of the society and its waste administration framework. Therefore, real issues of littering, waste dumping and other destructive effects on general health and the earth have risen. These issues now shape the fundamental foundation behind the present solid waste issues in developing nations.

Characteristics of solid waste problems in developing countries

Today, the whole populace of developing nations represents over 70% of the total populace. Waste administration in these nations is of grave worry from two perspectives. Right off the bat, the procedure of urbanization and populace fixation that is inseparably connected to waste administration issues is advancing at a pace that is substantially speedier than was at any point experienced by the present industrialized nations. The issue of waste administration in developing nations, along these lines, has risen as a critical and looming fiasco. Furthermore, these nations have frequent trouble in streamlining the institutional frameworks, managerial bodies, administration abilities and human assets that are expected to lead the pack in taking care of solid waste issues. It is in this way extremely troublesome for them to react adequately to the recently rising test of solid waste administration. The underdeveloped limit of the applicable performing artists in developing nations has been called attention to by numerous givers in the field of worldwide participation since the 1990s. This is a circumstance that is basic to many developing nations because of the negative heritage of long stretches of pioneer subordination, training issues and different elements.

Developing nations are in no way unconscious of the ecological contamination and solid waste issues that industrialized nations have encountered. In developing nations, as in industrialized nations, open mindfulness with respect to the requirement for enhancements in ecological quality is developing, and the significance of legitimate waste administration is winding up more broadly perceived. This developing mindfulness is reflected in the quantity of solicitations for development help or technical collaboration in the field of waste administration. Be that as it may, as talked about above, legitimate waste administration can be accomplished just through thought of its nearby association with the condition of the society, which is thus in view of the condition of its economy and establishments, and in addition its history and culture. The simplistic transfer of waste administration innovations, or the transplanting of waste administration frameworks from industrialized nations, can be useless. The technical writing distributed in industrialized nations generally manages advancements and frameworks that are excessively costly or too innovatively refined, making it impossible to be brought into any developing nation. This writing can't be utilized as it is even as materials for instructional classes (Flintoff, 1984) . The presentation of the encounters and technology of industrialized nations without respect for the conditions in the recipient nation has in a significant number of cases wound up as the shocking burden of an improper framework. In synopsis, there is really a certain, exceptional and settled foundation to each solid waste issue. It is recently difficult to locate an all-inclusive panacea.

External and Internal Problems

There are, in any case, ordinary phenomenon's and issues in regards to the solid waste issues confronting many developing nations, and it merits condensing them with a view to clearing up the way to manage singular solid waste issues. In light of this point of view, the predecessor report

(Institute for International Cooperation, JICA, 1993) by this Study Group characterized for illustrative purposes the waste administration issues regularly found among developing nations into three gatherings, as either 'external,' 'all the while external and internal,' or 'extraordinarily inside,' from the perspective of the municipal waste administration sector. The issues in this way arranged are summarized in Table 4.

Table 4: Waste Management Problems Categories in Developing Countries

(1) External problems	• population explosion, rapid urbanization, expansion of squatter settlements • socio-economic crisis • insufficient public education and community participation
(2) Simultaneously external and internal	• rapid increase in the volume of municipal and industrial solid wastes, lack of waste reduction efforts • lack of interest in solid waste problems among central and local government authorities • immature system of local autonomy • lack of coordination among sectors, organizations and municipalities • lack of a clear policy on waste pickers • lack of capacity in public cleansing departments, inadequate development programs for human resources • friction between management and labor within public cleansing departments or services • inadequate legal systems, insufficient law enforcement • weak financial base
(3) Internal problems	• lack of organizational capacity in municipal waste management • lack of short-, medium- or long-term planning • inadequate operation and maintenance structure for machinery and equipment, low utilization rate • use of technology that is technically, economically or socially inadequate

Source: YOSHIDA Mitsuo from (Sakurai, 2000)

The expression "external" here can be considered as alluding to issues that are caused by significant social changes that are firmly identified with the upstream production and utilization exercises from the perspective of the "metabolism of human society in connection to the utilization of materials" said above. Concerning the expression "inside," this can be considered as alluding to the downstream disbenefits and mismatches that happen because of social changes.

Impeding the benefiting process from assistance

A return of various past instances of development collaboration raises different issues and difficulties, which we to a great extent get from human, technical, organizational, financial, institutional, social and monetary constraints in favor of both the contributor and the recipient (Sakurai, 2000). These constraints in the developing nations and giver nations are talked about below:

Constraints faced by developing countries

Technical factors

In most developing nations, there commonly is an absence of human assets at both the national and local levels with technical expertise obligatory for solid waste management arranging and operation. Many officers responsible for solid waste management, particularly at the local level, have practically zero technical foundation or preparing in building or management. Without a sufficiently prepared work force, an undertaking started by external consultants couldn't be proceeded. Accordingly, the development of human assets in the recipient nation of external help is fundamental for the sustainability of the collaborative undertaking.

Another technical constraint in developing nations is the absence of general plans for solid waste management at the local and national levels. Subsequently, a solid waste technology is frequently chosen without due thought to its appropriateness in the general solid waste management framework. Now and again, foreign assistance is given to a component of a solid waste management framework for which the utilization of assets may not be most right. For example, an external help organization gave its help to enhance a general disposal site. In any case, the coverage of solid waste accumulation benefit is low to the point that solid waste generated is dumped at numerous undesignated sites (e.g., open territories, water channels, lanes, and so on.). Subsequently, improving the disposal site, in spite of the fact that it may not be a terrible venture, would have little impact on the general solid waste management adequacy. In such a case, the low accumulation coverage is a bottleneck in the general solid waste management framework in the city, and it would be most financially right to give assets to redesign the gathering administration.

Innovative work exercises in solid waste management are regularly a low priority in developing nations. The absence of innovative work exercises in developing nations prompts the choice of inappropriate technology as far as the local climatic and physical conditions, financial and human asset capacities, and social or cultural agreeableness. Subsequently, the technology chose can never be utilized, squandering the assets spent and making the undertaking unsustainable. A few aides/ manuals on proper solid waste management advances in developing nations are accessible in the literature, and the choice of technology could be made now and again in view of these aides/manuals. In any case, by and large, these aides/manuals must be changed to the local conditions prevailing in the nation, and consequently local investigations are ordinarily still required. Such examinations can be moderately effectively incorporated into a collaborative task and, to the degree conceivable, ought to include local research establishments.

Human Resources Factors

Human assets for waste administration are rare or, now and again, absolutely inexistent. Authorities without technical preparing in waste administration are doled out to deal with the difficult issues included, and the premise on which they may then get technical assistance can in this way additionally be constrained. This absence of limit as to the human assets accessible to oversee solid waste is likewise an explanation behind the absence of complete waste administration arranging in developing nations. Moreover, the gathering and examination of solid waste information are for the most part not given adequate consideration. Subsequently, there are a couple of chances for waste administration executives to end up specialists and to detail and actualize waste administration designs that are custom-made to the genuine circumstance in their nation. This thus makes it to a great degree hard to permit or develop advances that are most appropriate to the neighborhood conditions.

Financial factors

When all is said in done, solid waste management is given a low priority in developing nations, with the exception of maybe capital and extensive urban communities. Subsequently, extremely constrained funds are given to the solid waste management sector by the governments, and the levels of services required for security of public health and ecology are not achieved.

The issue is acute at the local government level where the local tax collection framework is insufficiently developed and, subsequently, the financial reason for public services, including solid waste management, is frail. This powerless financial premise of local governments can be supplemented by the accumulation of client service charges. Be that as it may, clients' capacity to pay for the services is extremely constrained in poorer developing nations, and their eagerness to pay for the services which are irregular and inadequate isn't high either. A compelling strategy for raising funds should be looked in any collaborative venture to ensure its sustainability.

Notwithstanding the constrained funds, numerous local governments in developing nations need great financial management and arranging. For example, in a town in a developing nation, more than 90% of the annual budget used for solid waste management was spent inside the initial half of the year. The absence of financial management and arranging, particularly cost bookkeeping, exhausts the constrained assets accessible for the sector much more rapidly, and causes the solid waste management services to stop for a few periods, accordingly losing the trust of service clients.

Waste administration services are by and large a low-priority situation in government budget allocations, along these lines the financial base for these exercises is frail. This is particularly valid for nearby governments who are the genuine regulators of solid waste administration programs. To compensate for deficiencies in the budget allocations, municipalities have tended to change from aggregate municipal junk disposal to outsourcing contracted administrations. In any case, in developing nations, there is a wide dissimilarity in the capacity of occupants to pay client fees for gathering garbage, and thus the municipal fiscal circumstance has regularly barely made strides. This represents a test for those associated with attempting to establish sustainable waste administration frameworks. Regardless of the possibility that the underlying venture costs are secured by development assistance from benefactors, a few times the in the past allotted funds are then regularly required for the operation and maintenance (O/M) of the solid waste administration system8. Furthermore, many developing nations experience the ill effects of an absence of limit in fiscal arranging and administration. Thus, waste administration programs stay unsustainable and open trust is lost.

Organizational, structural and institutional factors

A few organizations at the national level are typically required in any event partially in solid waste management. In any case, there are frequently no reasonable parts/functions of the different national organizations characterized in connection to solid waste management and furthermore no single office or advisory group assigned to facilitate their undertakings and exercises. The absence of coordination among the applicable offices regularly brings about various offices turning into the national counterpart to various external help organizations for various solid waste management collaborative tasks without monitoring what other national offices are doing. This prompts duplication of endeavors, squandering of assets, and unsustainability of general solid waste management programs.

The absence of proper legislation for solid waste management, which is a norm in most developing nations, is partially in charge of the parts/functions of the pertinent national offices not being plainly characterized and the absence of coordination among them. Legislation identified with solid waste management in developing nations is generally fragmented, and a few laws (e.g., Public Health Act, Local Government Act, Environmental Protection Act, and so on.) incorporate a few clauses on rules/controls in regards to solid waste management. The standards and controls are enforced by the diverse offices. Be that as it may, there are frequent duplication of obligations of the organizations included and gaps/missing components in the regulatory provisions for the development of powerful solid waste management frameworks. It ought to be additionally noticed that legislation is just powerful on the off chance that it is enforced. Accordingly, thorough legislation, which maintains a strategic distance from the duplication of obligations, fills in the gaps of critical regulatory functions, and is enforceable is required for sustainable development of solid waste management frameworks.

On account of a low priority given to the sector, the institutional limit of local government offices associated with solid waste management is by and large frail, particularly in little urban communities and towns. Local ordinance/by-laws on solid waste management isn't likewise very much developed. These frail local government establishments are not given clear mandates and adequate assets to satisfy the mandates. In vast metropolitan zones where there is more than one local government, coordination among the local governments is critical to accomplish the most adapt alternatives for solid waste management in the territory. For example, the siting of a solid waste transfer station or disposal office for use by more than one local governments is financially appropriate because of its economy of scale. Be that as it may, as these facilities are typically viewed as undesirable establishments and make not-in-mypatio (NIMBY) disorders among the occupants, no local government will find them inside its limit. The absence of a planning body among the local governments regularly prompts disintegrated and unsustainable projects for solid waste management.

The waste administration in developing nations is from time to time integrated, and there is regularly no unmistakable task of obligations regarding errands and timetables among the associations included. Moreover, there is regularly no umbrella association to facilitate overlapping duties regarding waste administration that include more than one organization. This circumstance not just hinders the viable usage of waste administration operations, yet additionally delivers perplexity in connection to technical collaboration and development assistance ventures among the contributors. Alongside these organizational and structural issues, the absence of a compelling lawful framework and technical norms constitute a noteworthy constraint. By and large talking, there is no integrated lawful framework to manage waste administration in developing nations. What's more, legitimate arrangements identified with solid waste are regularly incorporated as fragmented components in disparate laws, for example, laws for open cleanliness, neighborhood organization, and condition assurance.

Economic factors

Monetary and industrial improvement assume main roles in solid waste administration. Obviously, an enhanced economy enables more funds to be allotted for solid waste administration, providing a more sustainable financial basis. In any case, by definition, developing countries have frail financial bases and, henceforth, insufficient funds for sustainable improvement of solid waste administration systems.

Nearby industry which produces generally inexpensive solid waste gear and vehicles will lessen, or in some cases could eliminate absolutely, the requirement for importing expensive outside hardware/vehicles and in this way foreign exchange. Such nearby industry can also supply associated spare parts, absence of which is regularly responsible for sporadic and insufficient solid waste gathering and disposal services. In any case, the absence of industry manufacturing solid waste gear and spare parts and a restricted outside exchange for importing such hardware/spare parts are the lead instead of exemption in developing countries. Also in small developing countries, waste recycling activities are influenced by the accessibility of industry to get and process reused materials. For instance, the recycling of waste paper is possible just when there is a paper mill within a distance for which the transportation of waste paper is economical. The frail industry base for recycling activities is a typical constraint for the change of solid waste administration in developing countries, such as those in the Pacific locale where a substantial volume of bundle waste is produced.

The level of monetary development is firmly connected to fiscal viewpoints, as well as to the suitability of legitimate waste administration. Financial development is indispensable as a confirmation of the establishment of a fiscal reason for waste administration, and mechanical development is vital for the procurement and maintenance of machinery and gear for waste administration. Furthermore, the development of close by recycling enterprises is a prerequisite for the recycling of solid wastes; recycling is not typically doable if high transportation costs are caused in conveying recyclable materials to recycling plants overseas or in remote ranges.

Social factors

The social status of waste administration laborers is for the most part low. Now and again laborers originate from a particular social stratum, or position, and in others the business is controlled by a 'waste clan.' In such circumstances, hardworking attitudes and employment execution have a tendency to stay incapable. The private sector base that should generally complement public sector administrations is likewise somewhat frail in developing nations. Against this foundation, arrangements are regularly sought after through the participation of occupants in partnership with the group and NGOs. Extensive ecological instruction is in any case indispensable for group partnership to be genuinely successful. Social issues, for example, joblessness and poverty, are inseparably connected to the nearness of waste pickers, expecting thoughtfulness regarding the social ramifications of waste administration arrangement development. In developing nations, be that as it may, interaction between the regulatory specialists and subjects is uncommon, and the open doors for partnership are remote.

Because of insufficient resources accessible in the administration sector, collaborative projects regularly have endeavored to prepare group resources and create group self-help activities. Results are a blend of success and failures. Failed projects with inactive communities usually did not give individuals in the group with financial as well as social incentives to take an interest in activities. The social incentive is based on the responsibility of individuals as a feature of the group for the change of the group, and is made by open awareness and school instruction programs. The absence of open awareness and school instruction about the significance of appropriate solid waste administration for wellbeing and prosperity of individuals severely restricts the use of group based approaches in developing countries.

At dump sites, transfer stations, and street refuse bins, waste picking or scavenging activities are basic scenes in developing countries. Individuals involved have not gotten school instruction and professional training to obtain learning and skills required for different jobs. They are also influenced by restricted business opportunities accessible in the formal sector. The existence of waste pickers/scavengers creates regularly an obstacle to the operation of solid waste accumulation and disposal services. In any case, if sorted out legitimately, their activities can be successfully incorporated into a waste recycling system. Such an opportunistic approach is required for sustainable improvement of solid waste administration programs in developing countries.

Constraints identified with Donor Countries

Technical factors

Industrialized countries, which give external support to developing countries, usually have specialized expertise and HR suitable for solid waste administration in these countries. Their school and university instruction and subsequent at work training are targeted for the technologies of solid waste administration applicable to these countries. In any case, there is the absence of HR with sufficient experiences and learning of solid waste administration in developing countries. Opportunities to learn solid waste administration problems and practices in developing countries through customary training programs and seminars are seldom given in industrialized countries.

The absence of information and involvement in solid waste administration situations in developing countries leads to an inclination to support and give the technologies accessible in the contributor nation regardless of their relevance to the developing nation situation. In some cases, the solid waste administration gear and facilities, which are obsolete and outdated in the benefactor nation, are given as outside guide to the recipient nation.

Correspondence between consultants given by the external support office and the nearby partner in the developing nation sometimes becomes a constraint in implementing a successful collaborative undertaking. The correspondence trouble occurs in two distinct situations: (I) no basic spoken language exists between the external consultants and the nearby partner; and (ii) the neighborhood partner does not understand specialized terms. Efforts by the two sides to enhance correspondence capacity are being made in various countries.

As said before, the absence of a general arrangement for solid waste administration leads to a solid waste administration system which is not cost-viable. It also encourages a piece-meal approach by the external support organization. Referring to the prior case of support for development of a disposal site, it can be easily seen that the external support organization settled on the decision to support without sufficient consideration to different components of solid waste administration. Piece-meal, or not comprehensive approaches taken by external support agencies, frequently result in unsustainable solid waste administration projects.

The aptitude and technologies that industrialized contributor nations can offer don't really coordinate the requirements of developing nations as far as the accessible human assets where waste administration is at a more basic stage. Therefore, there is regularly a critical gap in comprehension and correspondence between the technical specialists dispatched by the benefactor and the authorities of the recipient government.

Financial factors

Obviously, all contributor agencies have their highest limits to which they can give financial support. Solid waste administration is one of numerous sectors for which an external organization provides its resources. For some giver agencies, solid waste administration may not be a need sector for support. As a result, there is a finite (and frequently constrained) measure of funds that can be apportioned to the sector.

Because of its inherent nature, solid waste administration does not render itself to an operation which can easily produce revenues. This is especially valid in developing countries where the willingness and capacity to pay for solid waste administration services are low. For external lending agencies, this means the risk of providing a credit to such a task is for the most part high. The high risk of credit projects can be lessened by building into the projects income raising systems (e.g., user charges, sales of reused materials).

Investors have budgetary constraints of their own, regardless of the waste administration needs of developing nations.

Institutional and structural factors

External support agencies have their own particular authoritative mandates and structure that restrict their activities to certain operations such specialized collaboration, advance/lending of capital funds, training, et cetera. Indeed, even in the same giver nation, there are usually extraordinary external support agencies, each specializing in one region of support. The degree of their geographical scope is also restricted to certain countries for their support. These authoritative mandates and operational scope of external support agencies determine the levels and types of resources given to solid waste administration projects in developing countries.

As specified above, in many cases their support is piece-meal and not comprehensive as individual projects to be useful in introducing substantial and lasting impacts on solid waste administration in the recipient countries. There is also absence of coordination among the various external support agencies to supplement each other's efforts, despite the fact that it is step by step improving as of late. With better coordination and correspondence among them, the sustainability of solid waste administration projects in recipient countries will be made better.

Diverse organizations in the contributor governments have different levels of authority and varying obligations. The mix of different cooperation plans for waste administration ventures requires coordination among the disparate organizations included, making it to a great degree hard to shape a complete task bundle. Considerable exertion is along these lines used in coordination among the diverse associations previously achieving an accord on a general bundle including advances, technical cooperation and hardware supply.

Economic factors

The financial situation of the benefactor is a determinant to the measure of funds that can be allotted for outsiders help to developing countries. Thus, it influences the levels of resources gave to solid waste collaborative projects. In any case, the financial situation of one funding nation is not all that basic for the sustainability of solid waste administration projects in developing countries.

External support agencies in industrialized countries have a tendency to advance solid waste administration technologies created in their countries and use consultants from their countries. It is understood and frequently acknowledged that there is a bias in the selection of gear, facilities, and consultants for solid waste administration collaborative projects. As specified before, the provision of solid waste hardware was done from the point of perspective of the giver office, instead of the need of the recipient nation. For instance, two vast compactor trucks of 8-ton limit each were given to the capital town of a small island nation where an estimated 7 tons for each day of solid waste was produced and there were many restricted streets. In another developing nation where solid waste is wet and has a low calorific esteem, the construction of an incinerator was suggested by a gathering of consultants from a created nation where incineration is exceptionally normal. Frequently, the appropriateness of an innovation to be used in a developing nation is not completely assessed, and the innovation is adopted based on the standard and experience of the contributor nation.

Investors are regularly disposed to pick groups of consultants from their own particular nation, and, subsequently, wind up choosing techs and hardware that are not really appropriate to the necessities of the recipient nation. This is really one of the principle reasons that the advances that are subsequently transferred end up being wrong. To beat such limitations, the more extensive usage of neighborhood consultants is required.

Social factors

In any nation, created or developing, there are cultural or social norms acknowledged just by the society. Such norms influence designs of solid waste administration systems. Where the society allows just a certain social class or gathering to manage solid waste, the accessibility of work drive for solid waste accumulation and disposal becomes constrained by this run the show. In some countries, directly handling human waste is a customary taboo, which at that point prohibits the utilization of co-composting of refuse and human waste. The absence of understanding of nearby cultures and ways of life by the external support office is frequently a cause of disappointment of a collaborative venture.

Correspondence trouble was refered to as a constraint prior. Notwithstanding the dialect related correspondence issue, the absence of better than average disposition and experience of external consultants in working with officials of developing countries results in unnecessary tension between the consultants and nearby partner.

The restricted transfer of technology and undertaking arranging is the approach frequently taken by investors, with little respect for the distinctions in social frameworks, societies and traditions amongst them and the recipients.

Further development in waste management

This outline of different parts of waste administration in developing nations and the constraints to global assistance can be considered as the foundation to limit development. As was underlined in the predecessor report (Institute for International Cooperation, JICA, 1993), the principal thought that waste administration specialists from industrialized nations need to see in any technical cooperation venture is to follow the waste administration techniques that have evolved historically in the group and that are right now being honed in that society, and to break down the foundations of these strategies in connection to the condition of the society and the idea of the challenges that have emerged from late fast social changes. On the off chance that this approach is taken, the genuine undertaking of waste administration specialists from industrialized nations is to assist the recipient nations with developing their ability to conquer these troubles and to establish their own particular new waste administration frameworks that react to the real circumstance they confront, thinking about the procedures through which the previous strategies have been obtained and streamlined throughout the years. At the end of the day, the errand from now on is to stretch out assistance to help limit development in waste administration.

In the main paragraphs of this section, it was said that waste resembles a mirror that reflects different parts of a society. To put it in an unexpected way, solid waste issues can't be explained autonomously from different issues; an approach that encompasses different elements, particularly societal, is a flat out prerequisite. This is very obvious from the historical advancement of Japan's waste administration operations and the present circumstance as it was depicted in past sections. The on-screen characters associated with limit development can be classified into three levels: people, associations and establishments/society 10. It ought to be reiterated here that for the determination of solid waste issues, the accessible limit at the institutional and societal level is an essential component, notwithstanding the need to reinforce the limit of the people and associations engaged with the waste administration sector, who are the conventional focuses of technical cooperation.

Recent Trends in official development assistance for waste management

Since the 1970s numerous investors have broadened multilateral and two-sided development assistance in the solid waste administration sector to react to issues emerging from the rapid advance of urbanization in developing nations. In the paragraphs that follow, a review is given of the lessons gained from the genuine encounter of these contributors through their execution of solid waste ventures. The most recent patterns that have evolved from reflection on the lessons gained from overseas development assistance and cooperation in the field of waste administration are then reviewed.

Demands from developing nations for development assistance and cooperation in the field of waste administration have been made with particular emphasis on ventures in the capital city and other urban regions where there has been a rapid ascent in the convergence of the populace, and benefactors have reacted with many allows or credit ventures. The methodologies of these ordinary development assistance and cooperation programs for urban waste issues in developing nations can be arranged into the following three classifications.

Hardware Approach

The underlying methodology of overseas development assistance in such manner depends on the arrangement of machinery and gear, for example, waste accumulation vehicles, gathering gear, handling facilities and other hardware, with a specific end goal to strengthen the current accumulation fleet as part of urban natural change endeavors. This is the approach most of the time utilized since the 1970s, which was a time of new activities in overseas development assistance and cooperation in the solid waste administration sector. A regular case is the urban development program that the World Bank advanced from 1976 of every 40 urban areas around the globe, including Cairo, Alexandria, Manila, Djakarta and Singapore (Cointreau, 1982). A sum of more than $500 million was invested into this program for waste administration alone. Nonetheless, these funds were not coordinated to free waste administration ventures, but rather were visualized as part of general urban change extends that additionally included open water supply, sewage, open transportation, et cetera. As a rule, overseas development assistance regularly appeared as giving waste gathering machinery or waste disposal facilities (Bartone, 1990).

At times, such hardware-based ventures contributed to raising the waste accumulation rates in the urban areas included and to the nonstop change of urban waste administration operations. A notable model of good practice is the World Bank venture in Singapore. Under strict control by the recipient government, a fundamental change of the waste administration framework, including the development of new incineration facilities, was done and another framework for the nation's waste administration operations was established. Likewise, it was around this period that manuals were composed on the philosophies for gear supply and other technical cooperation/aid extends and proper waste administration advances, drawing upon the numerous genuine cases of overseas development assistance in waste administration (Cointreau, 1982).

Subsequent assessments have uncovered that however much hardware may have been given, this sort of development assistance just has a brief, unsustainable impact and could even have unfavorable outcomes in situations where the recipient developing nation does not have a ground breaking strategy (M/P) or medium-to long haul arranging process for waste administration, or does not have the nonphysical (operational) methods and knowledge to use and keep up the gear and facilities that are given, or has an outmoded authoritative framework (Bartone, 1990).

Software Approach

Reflection on the principal kind of approach as specified above prompted the presentation of another approach in light of development assistance for non-physical angles, for example, arranging, operations, maintenance and organization. In this report, this approach is characterized as the arrangement of assistance for arranging, the development of waste administration frameworks and technology transfer that spotlights on the change of essential mechanical limit. Mr. Carl Bartone, a urban arranging pro who was for some time associated with waste administration ventures at the World Bank, has emphatically pushed the requirement for assistance to the arranging viewpoint, in light of his experience of the World Bank's undertakings in the 1980s, as depicted over 16. As indicated by this approach, the indispensable components were the lead of preliminary investigations going before the genuine contribution of hardware adopted under the principal strategy and the arrangement of waste administration designs (nonphysical sources of info). These investigations included reviews of the real solid waste circumstance, M/P planning, administration and financial examinations, feasibility thinks about (F/S) and outline angles. As will be found in the following section, this 'product based' approach was really epitomized in the numerous development studies and give aid extends that JICA stretched out in the solid waste administration sector from the finish of the 1980s through the 1990s.

The center of this approach is not the straightforward contribution of hardware, but instead the arrangement of a M/P encompassing each waste administration part (contribution of 'programming'). The contribution of vehicles, gear and other hardware is completed just amid the way toward working up the waste administration framework as per the M/P. Regardless of whether a sustainable waste administration framework in light of preliminary investigations and the subsequent arrangement is effectively developed, in other words, whether the sources of info and development assistance in the non-physical field are viably made use of, generally relies upon whether if there is an open domain in the recipient nation as far as the accessible human assets, association, fund, structure and foundations. As it were, regardless of whether there is both the capacity to actualize (limit) and free initiative (possession). At the point when the executing organization has a specific level of limit, this progression (development assistance) works as the impetus for development assistance recipients to take their own drive to enhance and establish a waste administration framework, despite the fact that there may at present be different barriers to this. Cases of this incorporate the waste administration frameworks of Bangkok, São Paulo, Mexico City, Almaty, Kazakhstan, and Panama that had a moderately abnormal state of human assets limit and organizational structure yet likewise had a feeling of urgency concerning waste issues because of the rapid development of their economies. Be that as it may, where the limit or proprietorship is not adequate, the M/P can't be implemented and the recently developed waste administration framework and the provided machinery and hardware can't be maintained. A significantly more risky perspective is that any failure could bring about undue dependence on overseas development assistance.

A case of a distributed critical investigation composed from the standpoint of developing nations on fizzled cases is the one on Katmandu, Nepal (Thapa, 1998). For a time of over 12 years from the mid-1980s to the 1990s, the city got from a bilateral contributor persistent development assistance under a technical cooperation plot. The arranging, organizational structure setup, machinery and gear supply, technology transfer, human assets preparing, pilot venture execution and every single other action possible at the time were all systematically given to the city. In any case, quickly after the benefactor pulled back from the city in 1993, the waste administration framework winded up plainly dysfunctional, and thc waste issues moved toward becoming exacerbated to a similar degree that urbanization had advanced. For this situation, the purpose behind this weaknesses in the regulatory body and framework, is an underdeveloped managerial limit concerning operations and back, and a couple of chances for participation by the occupants. Undoubtedly, numerous non-physical information sources were given for this situation, notwithstanding the hardware. Nonetheless, underlying this disastrous failure was a deficiency in organizational, institutional and structural limit, stemming from the feeble feeling of possession. To the extent development assistance with respect to organizational, institutional and structural angles are concerned, any investor, who is after an outsider, can be 'locked in' in broadening circuitous help, however can never 'give' such assistance all things considered. This is illustrated well by the Katmandu case in which the framework wound up noticeably dysfunctional promptly after the venture was finished.

Capacity Development Approach

The third approach speaks to another heading in development assistance that depends on reflections on involvement with the second approach and positions limit development of the principle on-screen characters at the bleeding edge of waste administration in the developing nation. This approach is characterized here as a procedure of general cognizance and assessment of the current establishments and arrangement condition, and the level of limit of the recipient nation as indicated by its social and social foundation, and an examination of the most fitting substance, strategy and technique of the assistance. At the end of the day, it is an approach in which the initial step is to help enhance the limit and afterward to develop the physical and non-physical sorts of assistance completely and as per the recipient's subsequent limit.

In the good 'old days, the 'performing artists' associated with limit development were believed to be nearby government authorities who were accountable for waste administration operations and the emphasis was on the development and change of these human assets, and in addition the association and administration of managerial bodies in charge of waste administration operations, which means customary 'institutional development' in the narrow sense. As far as organizational fortification, for instance, the technique for situating waste administration as part of general urban change extends as was seen in World Bank extends in the 1980s was acquired by the Healthy City Projects (HCPs) that WHO/ UNDP advanced amid the 1990s in urban areas in Asia, Africa and Latin America. These tasks had the point of redesigning and renewing individual urban natural operations through a cross-sectoral approach fixated on open cleanliness and general health ventures, and, through this procedure, additionally had the point of enhancing the association and administration of neighborhood managerial bodies. In this unique situation, such undertakings can be viewed as a type of 'institutional development' assistance in the narrow feeling of the term (Harpham, Burton, & Blue, 2001).

Limit development, in the narrow idea of 'institutional development' in connection to waste administration experts, which are for the most part nearby government organizations as portrayed above, is in itself a critical assignment. In any case, in all actuality, waste issues are hard to explain through limit changes in authoritative bodies alone. It has progressively been brought up that there is a requirement for another approach from the societal perspective accentuating agreement building, group participation and partnership. These references to social partnerships originate from the following circumstance. Right off the bat, the proceeding with deterioration in the earth because of the escalation of populace fixations and rapid urbanization and development has overwhelmed the capacity of nearby governments to react. Besides, under such circumstances, groups have needed to turn for assistance to CBOs and NGOs with a specific end goal to address the issues. Thirdly, subsequently, the connection between the regulatory experts and the nearby groups has changed, and it has turned out to be generally perceived that groups themselves come to assume an indispensable part in urban ecological administration. What's more, the ideas of 'group participation in solid waste administration' and 'group based solid waste administration (CBSWM)' are progressively being pushed as a methods for reacting to the deficient reactions from neighborhood governments (Moningka, 2000). This change is portrayed as the augmentation and extending of the scope of waste administration performers engaged with the answer for related issues confronting urban regions. Thus, the objectives of limit development must be stretched out to every one of the stakeholders included (the occupants and their groups, NGOs, CBOs and private modern companies), and the partnership between them has been reinforced, prompting those occupied with development assistance grasping an idea of 'limit development' that encompasses all sectors of the society.

In this third approach, contracting to private sector and privatization are effectively sought after under the principle of 'public private partnership (PPP).' One of the current attributes is that such activities are taken in numerous urban communities of developing nations also.

It is a result of the estimated benefits, for example, enhanced waste administration benefit through quality change in operation/maintenance, lessened cost, better natural insurance through enhanced technology, and infrastructure speculation with private-sector funds 30. There are an assortment of structures running from partial contracting to whole privatization, and in developing nations too, NGOs and CBOs as well as authentic waste businesses are rising 31. It is important that with the advance of PPP, free accumulation and disposal of rubbish has been changed to paid administrations to some degree, and is very regular in many developing nations today. Be that as it may, paid administrations take diverse structures and are principally parted into 4 frameworks: "particular rate," the expense increases as the waste volume increases in this framework, "pay-for-vast volume," no expense charged until the point that specific waste amount is reached in this framework, "fixed charge," a fixed expense paying little heed to waste volume in this framework and "near particular charge," waste charge is connected to metered power utilization in this framework. Beneficial outcomes of changing to paid administrations are said to be:

- reduction of waste era,
- promotion of recycling,
- more equitable cost sharing would turn into a subject of dialog in the event of "particular rate framework,"
- prevention of sullying by business-related urban wastes,
- securing income, and
- increase of individuals' mindfulness.
- Then again, concerns are raised, for example,
- double tax collection,
- regressive to income level,
- increase of unlawful dumping, and
- not helpful for changing the monetary structure of large scale manufacturing/mass-utilization (Yoshida, 1998).

Further, private-sector participation could turn into a hotbed for defilement and social issues around personal stakes or entrance of 'waste mafia' that submit illicit dumping to avoid landfill costs, unless it is implemented legally and with adequate straightforwardness. In this manner, development of privately owned businesses by free rivalry, appropriate arranging, workplace, and observing and control of administrations are keys to the accomplishment of PPP. In this sense, change of initiative and managerial limit of the administration experts in solid waste administration is dependably an issue if private-sector contracting might be advanced. On account of tasks in Accra, Ghana and Hyderabad, India, privatization was presented at the request of the giver, despite the fact that the limit with respect to authoritative administrators had not been developed completely. While solid waste gathering was without a doubt enhanced, negative outcomes, for example, provincial partition, deteriorated workplace and increased natural effects were additionally detailed. There is a report that privatization was attempted as a handy solution to solid waste issues in the nearby urban communities of Helauda and Biratnagar, Nepal, however the private element needed to pull back in view of the absence of encounters and know-how in legally binding functioning relationship, leaving a stunned solid waste administration framework behind. Every one of these encounters recommend that privatization can cover just a bit of the limit required for solid waste administration and there is dependably a part to be played by the neighborhood government. Privatization can never assume control everything.

Likewise, a standout amongst the latest aid patterns brings the sexual orientation point of view into limit advancement. In solid waste administration, particularly in developing nations, ladies assume a conclusive part. Waste administration in family units is fundamentally a lady's activity in these nations. Many waste administration CBOs are sorted out by ladies (cases in Karachi, Chennai, Dacca, Hanoi). A lion's share of waste pickers are ladies and kids (case of Bangalore, India). The vast majority of sophisticated arranging is finished by low-wage female laborers (Bangladesh). The sexual orientation viewpoint will be an indispensable point of view in future social approach.

TRENDS OF SOLID WASTE MANAGEMENT FOREIGN AID

International Organizations and Agencies

World Bank and Regional Development Banks

The World Bank is the world's biggest remote aid association in solid waste administration with a 30-in number staff of municipal waste administration specialists. Since the 1980s the Bank has advanced more than 100 solid waste administration ventures worth a total of more than US$1 billion in developing nations. These ventures were done from fundamentally two methodologies:

- assistance in solid waste administration change from the perspective of urban development, and
- solid waste administration and disposal from the perspective of ecological insurance, contamination reduction, and expulsion of ecological defilement.

These methodologies appeared as projects; the previous approach is known as the Urban Development Program (UDP), and the last is incorporated into the New Ideas in Pollution Regulation (NIPR) as tasks for hazardous waste disposal. Be that as it may, taken through and through, the previous or urban development approach has by a wide margin the bigger number of tasks, demonstrating the predominantly solid propensity to take up municipal waste administration as one component of assistance to city administration change. The Bank considers that the key issues are strategic administration arranging, institutional courses of action, private sector participation, financial administration, and ecological administration. All the more as of late, there have been World Bank assistance extends particularly coordinated to solid waste administration, however not very numerous up until this point. Cases include: Toshkent (1998), Bosnia and Herzegovina (2002), Latvia (1999), and Mexico (1994).

Furthermore, development of cleaning service or solid waste administration is recorded as a high priority assignment for specific nations, for example, Ghana in the Bank's Poverty Reduction Strategy

(PRS), from the perspective of helping the battle against poverty issues in urban areas and giving city administrations to poor urban inhabitants. Like the World Bank, the Inter-American Development Bank (IDB) expands assistances for urban solid waste administration to The Bahamas and other Latin American nations, as a segment of urban development. IDB has obviously demonstrated its favored assistance to privatization ventures. Then again, the Asian Development Bank (ADB) has not really reported an arrangement of supporting solid waste administration in essence, yet like the World Bank, it has stretched out assistance to ventures that incorporates solid waste administration as a part of urban development. Additionally, ADB frequently manages the issues of waste pickers, squatters and other urban poverty issues through assistance to solid waste administration. Between city and provincial cooperation in solid waste administration is much of the time pushed. The European Bank for Reconstruction and Development (EBRD) is effectively occupied with various sewage/drainage and municipal solid waste administration change extends in the previous Soviet nation urban areas, incorporating Almaty and Toshkent in Central Europe, Budapest and Zagreb in Eastern Europe and St. Petersburg in Russia. Together with the change of obsolete facilities and frameworks, EBRD has adopted the strategy of pushing for privatization. The African Development Bank (AfDB) has few records of assistance in the solid waste administration sector.

United Nations Organizations

Joined Nations associations that render assistance to solid waste administration in developing nations are World Health Organization (WHO) and United Nations Environment Programme (UNEP). What's more, in partnership with these associations, United Nations Development Programme (UNDP) likewise assumes a part in solid waste administration. UN associations are not just included straightforwardly in the operation of technical cooperation and assistance programs yet in addition arranging an extensive part of their exercises to a whole and generalization of data, issuance of rules, and readiness of papers and manuals. They assume a critical part as the wellspring of universal data and information. Similarly vital is their part of proposing measures and rules identifying with solid waste administration, in light of their speculation work of data and information. It is in this way important to seek through the data made accessible by these associations, when one intends to render assistance to solid waste administration in developing nations.

WHO stretches out assistance to solid waste administration from the perspective of guaranteeing open cleanliness and healthy way of life of citizenry, and actualizes technical cooperation extends for the most part for disposal of medical and hazardous waste. Particularly in the field of irresistible medical wastes, WHO has prepared and distributed the most systematic manuals accessible today and has gathered the fitting advances. It is the hotspot for 'must read' materials for everybody advancing technical cooperation in medical waste administration. (WHO, 1999) UNEP, through its International Environmental Technology Center (IETC) situated in Osaka and Shiga, Japan, is occupied with an extensive gathering of technical data and collection of proper advancements for sewage treatment and solid waste administration from the perspective of urban condition administration. In partnership with UNEP, United Nations Industrial Development Organization (UNIDO) stretches out assistance to solid waste administration concentrating on African urban areas from their bases in Dares-Salaam and Nairobi.

Joined Nations Institute for Training and Research (UNITAR), in partnership with International Labor Organization (ILO), advances 'Synthetic and Waste Management Program' and 'Preparing and Capacity Building Program for Implementation of the Globally Harmonized System of Classification and Labeling of Chemicals (GHS). It has aggregated data on the administration of hazardous substance wastes and issues in workplace. On its part, ILO executes its own particular ventures of solid waste administration privatization with an end goal to grow work opportunity. Fruitful cases are accounted for in Dares-Salaam and different areas. Joined Nations Center for Human Settlements (UNCHS-Habitat) has established Urban Management Program, under which it has implemented assistance ventures for municipal solid waste administration from the standpoint of enhancing human habitat. Since 1991, it is occupied with Sustainable Waste Management in African Cities (SWMAC) that fundamentally targets urban areas in Africa. UNDP has advanced numerous solid waste administration extends in developing nations in cooperation with the World Bank, UNIDO, UNEP and UNCHS-Habitat. Late activities significant include: control era from landfill gas in light of the idea of 'waste-to-vitality', and change of landfill site (methane gas recuperation) in accordance with the scheme of clean development mechanism (CDM) for ozone harming substance lessening. UNDP has many encounters of working in partnership with bilateral benefactors including the Netherlands, Norway, Germany and the U.S. JICA's people group based solid waste administration venture (pilot venture) in Manila was succeeded and developed by UNDP in partnership with the Japanese government.

Bilateral Cooperation

Germany

Germany makes its bilateral cooperation through GTZ. Solid waste administration is viewed as a component of cooperation for condition and infrastructure, and three concentration territories have been recognized: hazardous waste, municipal waste administration and waste administration strategy. Like JICA, GTZ has implemented numerous solid waste administration extends in Asia, Africa and Latin America, and furthermore fills in as development advisor for particular activities. A trademark approach of GTZ is human systems administration. Systems are shaped between the GTZ specialists and the recipient nation's task individuals and counterparts, and data trade and follow-up are made. There are four such systems at display: Mediterranean Environmental Network - Near and Middle East, Maghreb (MEN - REM), Services on Water and Sanitation - Subsaharan Africa (SOWAS), Rural Development - Latin America and the Caribbean, and Transport, Environment, Energy, Water - Asia (TEEWAS).

The Netherlands

Technical cooperation in urban waste administration by the legislature of the Netherlands to developing nations is called Urban Waste Expertise Program (UWEP), and a non-benefit NGO situated in that nation by the name of WASTE is accused of the operational obligation. Numerous counsel sort limit development assistance ventures have been advanced instead of those of hardware input, and the Netherlands is at the front line in executing ventures in light of the group participation approach and thoughtfulness regarding sex issues. Reports and productions by WASTE have a high caliber as the essential data on solid waste administration cooperation to developing nations.

Switzerland

Swiss Agency for Development and Cooperation (SDC), as on account of the Dutch government, depends the operation of technical cooperation in condition, water and waste sector to a NGO situated in Switzerland called SKAT (Swiss Center for Development Cooperation in Technology and Management). It is particularly dynamic in sorting out classes and workshops and giving distribution and data administrations. Its cooperation ventures are for the most part limit development sort, rather than hardware input. It regularly works in partnership with WASTE or GTZ as for group participation and sexual orientation issues.

United States

Solid waste administration cooperation programs that the U.S. executes through U.S. Office for International Development (USAID) 60 are naturally implemented in various urban communities of an objective nation. Along these lines, a nation particular model of solid waste administration is planned one by one, capitalizing on the generally uniform common, social, social and financial states of the task areas. Target nations incorporate Egypt, India, Latin American nations and Eastern European nations. The principal approach is privatization of solid waste administration operation or contracting out to private sector. Notwithstanding the nations specified above, Norway, Denmark, Canada and France expand technical cooperation in solid waste administration. As of late, Europe's solid waste administration cooperation to Eastern Europe (local cooperation) is likewise made through European Union's EEA (European Environment Agency).

International NGOs

As we have found in the above with respect to bilateral benefactors the Netherlands and Switzerland-, there are NGOs that operate nearly in the bilateral cooperation scheme. Then again, there are a few NGOs that are assuming imperative parts universally in solid waste administration cooperation to developing nations. Universal Solid Waste Association (ISWA) is a global relationship with its headquarters in Denmark, and capacities as the worldwide union of academic social orders and open organizations of part nations. In the meantime, it has a character of a worldwide NGO. Support to waste administration in developing nations is plainly specified in its statement of purpose, and scientific data on solid waste administration is made accessible to developing nations. In particular, Waste Management and Research, ISWA's worldwide diary, and ISWA Newsletter regularly cover subjects identified with solid waste administration in developing nations, and ISWA distributes writings for developing nations.

The Water, Engineering and Development Center (WEDC)- an exploration and instruction focus of UK's Loughborough University- offers instructional classes in water supply administration, solid waste administration and sewage treatment and in addition postgraduate training and research. WEDC arranges a yearly meeting in a developing nation together with a neighborhood sorting out board of trustees framed by the nearby offices working in water supply, solid waste administration and additionally sewage treatment. The meeting procedures regularly contain articles and papers on solid waste administration in developing nations, and fill in as a global gathering for data trade that assistance explain the knowledge divide between industrialized nations and developing nations. The Collaborative Working Group for the Promotion of Solid Waste Management in Lowland Medium-Income Countries (CWG) is a system established in 1995 to support solid waste administration in developing nations. It was initially propelled by Switzerland's SKAT in conjunction with SDC's technical cooperation in solid waste administration. Be that as it may, CWG has evolved into a global system, as it continued holding yearly workshops in Switzerland, the U.S., Egypt and Brazil. The workshop gives a decent chance to data trade on proper innovations and strategy of cooperation/assistance.

KEY TO SUCCESSFUL COLLABORATION

The debates in the preceding sections indicate that there are various constraints looked by both the recipient and contributor countries in developing and implementing sustainable solid waste administration systems. There is no simple decision to expel or loosen any of these constraints. Some constraints are harder to evacuate than others. A large number of measures are usually required to create a successful outcome of a collaborative task. A blend of some of the following measures or approaches may prompt a successful outcome.

(a) Packaging External Support

Various external support agencies perceive solid waste administration as a need issue in developing countries and are interested in supporting to enhance the situation. In any case, their approaches to solving solid waste administration problems in developing countries have been piecemeal and not all around coordinated. Also, their support has been given mostly on a short-term basis. These characteristics of external support are inherent in the authoritative mandates and operational modes of the external support office, and in this way they can't be easily changed.

What can be changed, be that as it may, is to combine support from various international guide agencies to make a collaborative venture more comprehensive and long haul/continuous. This requires better coordination and correspondence among the external support agencies and improvement of partnership among them, removing the authoritative egos and sharing and contributing their resources to the benefits of the recipient nation. The collaborative task should be designed to enhance the solid waste administration situation step by step finished a long stretch, instead of attempting a handy solution.

(b) Defining Clear Roles of Relevant Agencies in Developing Countries

Better coordination for compelling usage of a solid waste administration collaborative venture is also required by the various agencies involved in solid waste administration in the recipient nation. Notwithstanding, as specified prior, numerous solid waste administration projects in developing countries suffer from the absence of coordination among the applicable agencies, which regularly results from the absence of clear roles defined for these agencies in solid waste administration. To ensure powerful institutional support for a collaborative undertaking for solid waste administration, the roles and responsibilities of the various agencies involved should be defined clearly and a coordination mechanism be established. This should be possible without drafting new legislation or amending the existing one, which is ordinarily a tedious exercise in any nation. A working gathering involving officials from the various agencies can be set up to discuss initially the roles and responsibilities of their respective agencies, and the working gathering can be later moved up to an administrative advisory group or task drive.

(c) Developing Human Resources

For sustainable solid waste administration in developing countries, human resource advancement should always be a piece of the external support bundle. Without neighborhood HR, a collaborative undertaking initiated by external support won't have the capacity to continue. To create HR with specialized expertise in solid waste administration in developing countries, there are three strategically critical groups for external support, to be specific (I) key personnel in the national coordinating unit of the focal government; (ii) operational managers of selected nearby governments; and (iii) university and other higher instructive institutions. Among these objective groups, the strengthening of HR in the national coordinating unit and maybe a couple selected nearby governments is the first need and should be done in short term while support to higher instructive institutions is a long haul program.

It is the responsibility of the recipient nation to select these potential target groups and enhance their correspondence abilities. The benefactor countries should also enhance their HR in terms of their correspondence capacity and information of solid waste situation in developing countries.

(d) Supporting Strategic Planning and Follow-up Implementation

General solid waste administration plans at both the national and nearby levels are essential for utilizing constrained resources most successfully, and providing an edge of reference for potential external support. Consequently, the formulation of national and nearby strategic plans for solid waste administration should be considered at the initial stage of the external support bundle.

Realizing the significance of strategic planning, numerous external support agencies have supported or are beginning to support the readiness of general national and nearby plans previously providing gear and facilities. In any case, the operation of such specialized assistance is regularly separated from that of the provision of loans and grants for facilities and gear. As a result, the subsequent activity to the planning assistance (i.e., provision of grants and loans for facilities and gear) is postponed or not given by any stretch of the imagination. Consequently, there are many plans created, yet they have not been actualized. For the sustainability of a solid waste collaborative task, it is critical to give external support to catch up on the execution of the arrangement arranged. Here again, the approach of packaging external support can assume a key part.

(e) Developing Self-financing Schemes

The governments of developing countries have restricted funds for solid waste administration and must create measures to diminish and recuperated the use and increase revenues where possible. They have to turn their solid waste administration systems to more self-financing programs. External support can be successfully used to create various option cost-cutting, cost-recovering, and income raising schemes (e.g., waste minimization, deposit-discount system for recyclable materials, import or sales assess on certain bundled products, gathering of user service charges, and so on.) and execute pilot studies on these monetary incentive measures.

Private sector support in solid waste administration gathering and disposal services is also an approach to lessen the financial weight of the legislature. It can draw not just investment finance from privately owned businesses for solid waste administration gear and facilities, yet in addition administrative expertise and specialized skills. Experiences in developing countries, which are accounted for elsewhere3, indicate that secretly worked services are for the most part more cost-successful than open sector services. In this manner, the use of private sector resources through an authoritative course of action provides a potential option towards self-financing solid waste administration.

Viable utilization of monetary incentive measures and private sector resources in solid waste administration requires HR to design and oversee such schemes. Aside from HR advancement in specialized aspects of solid waste administration, human resource improvement in financial planning and administration is necessary and frequently a key to the improvement of more self-financing schemes.

(f) Raising Awareness of the Public and Decision Makers

Viable administration of solid waste requires the participation of the overall population. Lifting the need of, and allocating more resources to, the solid waste administration sector needs the support from decision makers. It is, in this manner, vital to ensure that the society and decision makers' awareness activities are incorporated into the external support bundle. The point of these activities is typically long haul and it takes some force to develop before the effects are figured it out. Yet, once the interests of people in general and decision makers in improving solid waste administration are made, the sustainability of solid waste administration projects will be significantly better.

Enhanced awareness of decision makers may prompt changing national socio-monetary and industrial advancement policies and associated government programs for improving solid waste administration systems in developing countries. For instance, more financial guide and expense incentives might be introduced to energize the advancement of recycling industry and business, or laborer assurance programs might be given to enhance wages and working conditions of laborers, including solid waste administration workers. Changing national policies in contributor countries could also enhance ways in which their technologies are transferred to recipient countries.

REFERENCES

1. The Open University. 2017. *Solid Waste Reduction, Reuse and Recycling*. [ONLINE] Available at: http://www.open.edu/openlearncreate/oucontent/view.php?id=80574&printable=1. [Accessed 17 July 2017].
2. The European Commission. 2017. *Solid Waste Management*. [ONLINE] Available at:http://ec.europa.eu/echo/files/evaluation/watsan2005/annex_files/WEDC/es/ES07CD.pdf. [Accessed 17 July 2017].
3. Kryptone Consulting Limited. 2017. *Benefits of Proper Solid Waste Management*. [ONLINE] Available at:http://kryptoneconsultingltd.weebly.com/uploads/1/0/1/9/10194587/benefits_of_proper_waste_management.pdf. [Accessed 20 January 2014].
4. Rick LeBlanc. 2017. *An Introduction to Solid Waste Management*. [ONLINE] Available at:https://www.thebalance.com/an-introduction-to-solid-waste-management-2878102. [Accessed 18 May 2017].
5. Rick LeBlanc. 2017. *An Introduction to Solid Waste Management*. [ONLINE] Available at:https://www.thebalance.com/an-

introduction-to-solid-waste-management-2878102. [Accessed 18 May 2017].

6. GDRC. 2017. *Solid Waste Management*. [ONLINE] Available at: https://www.gdrc.org/uem/waste/swm-fogawa1.htm. [Accessed 17 July 2017].
7. Vamsi Krishna. 2017. *Treatment and Disposal of Municipal Solid Waste in Third world Countries: A Critical Review*. [ONLINE] Available at:http://www.academia.edu/8355947/Treatment_and_Disposal_of_Municipal_Solid_nWaste_in_Third_world_Countries_A_Critical_Review. [Accessed 17 July 2017].
8. Michaela Bobeck. 2010. *Organic Household Waste in Developing Countries*. [ONLINE] Available at:http://miun.diva-portal.org/smash/get/diva2:321886/FULLTEXT01.pdf. [Accessed 17 July 2017].
9. Charlott Durande. 2013. *Sustainable Waste Management Challenges in the South*. [ONLINE] Available at: http://ideas4development.org/en/sustainable-waste-management-challenges-in-developing-countries-charlotte-durand/. [Accessed 22 January 2013].
10. Alexander Fianko Otchere, Kwame Owusu Sarpong, Stephen Okyere. 2014. *Assessing the Challenges Affecting Solid Waste Management System in the Kumasi Metropolis*. [ONLINE] Available at:https://theartsjournal.org/index.php/site/article/view/323. [Accessed 22 May 2014].
11. European Parliamentary Research Service. 2015. *Understanding waste management Policy challenges and opportunities*. [ONLINE] Available at:http://www.europarl.europa.eu/RegData/etudes/BRIE/2015/559493/EPRS_BRI(2015)559493_EN.pdf. [Accessed 1 June 2015].
12. Urban waste management problems and challenges: the way forward. 2014. *Understanding waste management Policy challenges and opportunities*. [ONLINE] Available at:http://www.ghanaweb.com/GhanaHomePage/NewsArchive/Urban-waste-management-problems-and-challenges-the-way-forward-321845. [Accessed 20 August 2014].
13. My Speed Thesis. 2011. *Factors that Contribute for Poor Solid Waste Disposal Management in Ethiopia*. [ONLINE] Available at:

http://www.myspeedythesis.com/2011/03/factors-that-contribute-for-poor-solid-waste-disposal-management-in-ethiopia.html. [Accessed 15 March 2011].

14. Jessica McAllister. 2015. *Factors Influencing Solid-Waste Management in the Developing World*. [ONLINE] Available at: http://digitalcommons.usu.edu/cgi/viewcontent.cgi?article=1537&context=gradreports. [Accessed 2 November 2015].
15. Marianna Garfí. 2015. *Waste Disposal In Developing Countries And Emergency Situations. The Case Of Saharawi Refugees Camps.*. [ONLINE] Available at:http://www.iswa.org/uploads/tx_iswaknowledgebase/15-384paper_long.pdf. [Accessed 18 July 2017].
16. Chris Zurbrugg. 2010. *Basics of Solid Waste Management in Developing Countries*. [ONLINE] Available at:http://www.eawag.ch/fileadmin/Domain1/Abteilungen/sandec/publikationen/SWM/General_Overview/Zurbruegg_2002_SWM_DC.pdf. [Accessed 15 November 2010].
17. Waste Management. 2015. *aste Management in Northeastern Pennsylvania*. [ONLINE] Available at:http://alliancelandfill.wm.com/community/economic-benefits.jsp. [Accessed 18 July 2017].
18. A. Karagiannidis, St. Kontogiani. 2012. *Waste Management in Developing Countries*. [ONLINE] Available at: http://www.codwap.hs-bremen.de/02%20Material/HANDBOOK-%20WM_in_DC-CODWAP.pdf. [Accessed 5 March 2012].
19. Takele Tadesse. 2004. *Solid and Hazardous Waste Management.* [ONLINE] Available at:https://pdfs.semanticscholar.org/9abe/14b2c657a2295fade8a075cff9f008d07d03.pdf. [Accessed 4 August 2004].
20. Moritz Bühne. 2012. *3 Alternative Ideas for Waste Management in Developing Countriesr*. [ONLINE] Available at: https://www.ifu.com/knowtheflow/2012/3-alternative-ideas-for-waste-management-in-developing-countries/. [Accessed 9 November 2012].

21. Prokerala. 2017. *Tips to deal with plastic waste*. [ONLINE] Available at: http://www.prokerala.com/going-green/plastic-waste-management.htm. [Accessed 18 July 2017].
22. Javeriya Siddiqui, Govind Pandey. 2013. *A Review of Plastic Waste Management Strategies*. [ONLINE] Available at: http://www.isca.in/IJENS/Archive/v2/i12/14.ISCA-IRJEvS-2013-247.pdf. [Accessed 10 December 2013].
23. World Health Organization (1999):Safe Management of Wastes Ffrom Health Care Activities; A. Pruss, E.Giroult, P. Rushbrook
24. Gabre-Emanuel Teka (1997): Solid Waste Disposal From Food Premise; In Food Hygiene: Principles and Methods of Food-borne Disease Control with Special Reference to Ethiopia, Faculty of Medicine, Department of Community Health, Addis Ababa University.
25. Patricia A. Reagan and Jodi B. Fisher (1997): Community Health in The 21 Century; by Allyn and Bacon, A Viacom Company.
26. Monroe T. Morgan (1997): Environmental Health; Morton Publishing Company.
27. Green and Ottoson (1994): Community Health; 7th edition;
28. Tchobanoglous, Theisen and Vigil (1993): Integrated Solid Waste Managemen. Engineering Principles and Management Issues. McGraw-Hill, Inc.
29. Kebede Faris, Tssew Berhe, and Haile Mariam Churko (1993): Solid Waste Problem, Collection and Disposal Program in Jimma TowN, Ethiopian Journal of Health Development, Volume 7(2)
30. Sandra J. Cointrreau (1986): Environmental Management of Urban Solid Wastes in Developing Countries; A Project Guide. Solid Waste Management 111
31. Joseph A. Salvato (1982): Environmental Engineering and Sanitation; 3rd Edition.
32. P.Walton Pardon (1980): Environmental Health; 2nd Edition.
33. Frank Flintoff (1976): Management of Solid Wastes in Developing Countries; WHO

34. WHO, Regional Office for Europe. (1983): Management of Hazardous Waste; WHO Regional Publication, European Series No. 14. Copenhagen.
35. Ehlers, V.M (1965): Municipal and Rural Sanitation; 6th Edition, McGraw- Mill Book
36. Co., Inc.
37. Jordan's (1960): Tropical Hygiene and Sanitation; 3rd Edition by W.Wilkie
38. Hopewell, J., Dvorak, R., & Kosior, E. (2009). Plastics recycling: challenges and opportunities. *Philosophical Transactions of the Royal Society B: Biological Sciences*, *364*(1526), 2115-2126.
39. North, E. J., & Halden, R. U. (2013). Plastics and Environmental Health: The Road Ahead. *Reviews on Environmental Health*, *28*(1), 1–8.
40. Eerkes-Medrano D, Thompson RC, Aldridge DC Microplastics in freshwater systems: a review of the emerging threats, identification of knowledge gaps and prioritization of research needs. Water Res. 2015 May 15; 75:63-82. doi: 10.1016/j.watres.2015.02.012. Epub 2015 Feb 17.
41. Free CM, Jensen OP, Mason SA, Eriksen M, Williamson NJ, Boldgiv B High-levels of micro plastic pollution in a large, remote, mountain lake. Mar Pollut Bull. 2014 Aug 15;85 (1):156-63. doi: 10.1016/j.marpolbul.2014.06.001. Epub 2014 Jun 24.5.
42. Japan Ministry of the Environment, 2003: Fundamental plan for establishing a sound material-cycle society.
43. Japan Ministry of the Environment, 2006: State of discharge and treatment of municipal solid waste in FY 2004.
44. Kang JH, Kwon OY, Lee KW, Song YK, Shim WJ Marine neustonic microplastics around the southeastern coast of Korea. Mar Pollut Bull. 2015 May 12. pii: S0025-326X(15)00256-8. doi: 10.1016/j.marpolbul.2015.04.054.
45. Klein S, Worch E, Knepper TP Occurrence and Spatial Distribution of Microplastics in River Shore Sediments of the Rhine-Main Area in Germany. Environ Sci Technol. 2015 May 6.

46. Marine Plastic Pollution and Seafood Safety Zhonghua Liu Xing Bing Xue Za Zhi. Analysis on heavy metal pollution in major seafoods from Zhoushan Fishery, China]. 2012 Oct; 33(10):1001-4. Wang JY[1,] Wang YC, Lou JH.
47. Nate Seltenrich, New Link in the Food Chain? Environ Health Perspect. 2015 Feb; 123(2) A34-A41.Published online 2015 Feb 1. doi: 10.1289/ehp.123A34PMCID: PMC4314237 .
48. OECD, 2001: Extended producer responsibility: A guidance manual for governments. OECD publishers, Paris.
49. OECD, 2002a: Indicators to measure decoupling of environmental pressure from economic growth. OECD Environment Directorate. SG/SD (2002)1/FINAL, 108 pp.
50. OECD, 2002b: Environmental data waste compendium 2002. Environmental Performance and Information Division, OECD Environment Directorate. Working Group on Environmental Information and Outlooks. 27 pp.
51. OECD, 2003: OECD Environmental Data Compendium 2002. Paris. Accessed 25/06/07.
52. OECD, 2004: Towards waste prevention performance indicators. OECD Environment Directorate. Working Group on Waste Prevention and Recycling and Working Group on Environmental Information and Outlooks. 197 pp.
53. Ofira DD and Seneca JJ Biological effects and subsequent economic effects and losses from marine pollution and degradations in marine environments: Implications from the
54. Qureshi MS, Yusoff AR, Wirzal MD, Sirajuddin, Barek J, Afridi HI, Üstündag Z Methods for the determination of endocrine disrupting phthalate esters. Crit Rev Anal Chem. April 2015 Apr.
55. US EPA, Municipal solid waste generation, recycling and disposal in the United States: facts and figures for 2003. Washington, D.C., USA 2005.
56. US EPA, Global anthropogenic non-CO2 greenhouse gas emissions: 1990-2020. Office of Atmospheric Programs, Climate Change Division. 2006.
57. World Bank and IMF, Global Monitoring Report 2006: Ensuring

Environmental Sustainability Target 10. World Bank and International Monetary Fund, Washington, D.C. 2006.

58. WRAP 2006: Environmental benefits of recycling, an international review of life cycle comparison
59. Our Everyday Life. 2017. *The Advantages of Waste Disposal | Our Everyday Life*. [ONLINE] Available at: http://peopleof.oureverydaylife.com/advantages-waste-disposal-10100.html
60. http://www.johnsrefuse.com/blog/bid/227351/The-Importance-of-Industrial-Waste-Management
61. Mohit Iasija. 2017. *What are the useful methods of Hazardous Waste Management ?*. [ONLINE] Available at: http://www.preservearticles.com/201101082858/useful-methods-of-hazardous-waste-management.html
62. The Management of Hazardous Waste in Developing Countries | InTechOpen. 2017. *The Management of Hazardous Waste in Developing Countries | InTechOpen*. [ONLINE] Available at: https://www.intechopen.com/books/management-of-hazardous-wastes/the-management-of-hazardous-waste-in-developing-countries
63. The Management of Hazardous Waste in Developing Countries | InTechOpen. 2017. *The Management of Hazardous Waste in Developing Countries | InTechOpen*. [ONLINE] Available at: https://www.intechopen.com/books/management-of-hazardous-wastes/the-management-of-hazardous-waste-in-developing-countries
64. The Management of Hazardous Waste in Developing Countries | InTechOpen. 2017. *The Management of Hazardous Waste in Developing Countries | InTechOpen*. [ONLINE] Available at: https://www.intechopen.com/books/management-of-hazardous-wastes/the-management-of-hazardous-waste-in-developing-countries
65. The Management of Hazardous Waste in Developing Countries | InTechOpen. 2017. *The Management of Hazardous Waste in Developing Countries | InTechOpen*. [ONLINE] Available at: https://www.intechopen.com/books/management-of-hazardous-wastes/the-management-of-hazardous-waste-in-developing-countries
66. TreeHugger. 2017. *10 recycling and waste management trends to look out for in the near future : TreeHugger*. [ONLINE] Available at:

https://www.treehugger.com/corporate-responsibility/10-recycling-and-waste-management-trends-watch-out-near-future.html

67. Rethink Recycling. 2017. *8 Step Hazardous Waste Management Guide | Rethink Recycling*. [ONLINE] Available at: https://www.rethinkrecycling.com/businesses/waste-management-guide/hazardous-waste/8-step-hazardous-waste-management-guide
68. s for key materials in the UK recycling sector. Waste and Resources Action Program, Peer reviewed report prepared by H. Wenzel et al., Danish Technical University, published by WRAP, Banbury, Oxfordshire, England.2006.
69. Zinati, G.M., Y.C. Li, and H.H. Bryan, Utilization of compost increases organic carbon and its humin, humic, and fulvic acid fractions in calcareous soil. Compost Science & Utilization 9, pp. 156-162.2001.
70. AEA. (2001). *Waste Management Options and Climate Change.*
71. Bartone, C. (1990). World Bank's Cooperation for Developing Countries in the Solid Waste Sector: Their Experience, Implication Learned, and Future Strategy. *Journal of Japan Society of Waste Management Experts*, 2 (1), 59-65.
72. BBC. (2007). *East African ban on plastic bags.*
73. Bernstad, A., Jansen, J., & Aspegren, H. (2011). Property-close source separation of hazardous waste and waste electrical and electronic equipment e a Swedish case study. *Waste Manage*, 31, 536-543.
74. BMLFUW.(2015).*BundesministeriumfürLand-undForstwirtschaft, Umwelt und Wasserwirtschaft — Die Bestandsaufnahme der Abfallwirtschaft in Österreich.*
75. Chandra, H. (1999). *Environment News.*
76. Coffey, M., & Coad, A. (2010). *Collection of Municipal Solid Waste in Developing Countries.* Nairobi, Kenya: UN-Habitat.
77. Cointreau, S. (1982). *Environmental management of urban solid wastes in developing countries, A project guide.* Washington, D.C.: The World Bank.
78. Diaz, L., & Savage, G. (2003). *Risks and Costs Associated with Management of Infectious Wastes.* Manila, Philippines: WHO/

WPRO.

79. Diaz, L., Savage, G., & Golueke, C. (1996). Sustainable community systems: the role of integrated solid waste management. *Proceedings of 19th International Madison Waste Conference*, (pp. 280–291). WI, USA.
80. EEA. (2013). *Municipal Waste Management in France.*
81. EEA. (2015). *Waste prevention in Europe — The status in 2013.* European Environment Agency.
82. Eunomia Research and Consulting Ltd. (2002). *Costs for Municipal Waste Management in the EU. Final Report to Directorate General Environment. European Commission.*
83. Eurostat. (2011). *Municipal Waste Statistics in 2009.* Eurostat News Release 37/2011.
84. Flintoff, F. (1976). *Management of Solid Waste in Developing Countries.* New Delhi: World Health Organization.
85. Gupta, S. (2010). *Plague-like epidemic in Surat, India.* London, UK: Earthscan for UN-Habitat.
86. Harpham, T., Burton, S., & Blue, I. (2001). Healthy city projects in developing countries: the first evaluation. *Health Promotion International*, 16 (2), 111-125.
87. Hegde, V., Kulkarni, R. D., & Ajantha, G. S. (2007). *J. Oral Maxillofac. Pathol*, 11, 5-9.
88. Institute for International Cooperation, JICA. (1993). *Improving methods of waste management in the cities of developing countries.* Tokyo, Japan: Institute for International Cooperation, JICA.
89. Manaf, L., Samah, M., & Zukki, N. (2009). Municipal solid waste management in Malaysia: Practices and challenges. *Waste Management*, 29(11):2902-2906.
90. Mebratu, D. (1998). Sustainability and sustainable development: historical and conceptual review. *Environmental Impact Assessment Review*, 18(6): 493–520.
91. Mehdudia, S. (2009). *Making the most of waste: gold, power and more from Amsterdams refuse.* The Hindu.
92. MEJ. (2012). *SolidWaste Management and Recycling Technology of Japan.* Ministry of the Environment of Japan.

93. Moningka, L. (2000). Community participation in solid waste management, factors favouring the sustainability of community participation, A literature review. *UWEP Occasional paper*.

94. Mukesh, Y. (2001). *JK-Practitioner*, 8, 276–202.

95. Niedzwiecki, P., & Adams, A. (2007). *Household Hazardous Waste Report*. Massachusetts, USA: Cape Cod Commission.

96. Pasupathi, P., Sindhu, S., Ponnusha, B., & Ambika, A. (2011). *Int. J. Biol Med Res.*, 2(1), 472-486.

97. Patan, S., & Mathur, P. (2015). *Assessment of biomedical waste management in government hospital of Ajmer city – A study.*

98. Pathak, S. (1998). *Proceedings of National Workshop on Management and Hospital Waste*, 31-33.

99. Ponka, A., & Kaski, A. (1996). *Waste. Mange. Res.*, 14, 145–150.

100. Rajor, A., Xaxa, M., Mehta, R., & Kunal, J. (2012). *Environ Manage.*, 36- 41.

101. Rambol. (2006). *Waste to Energy in Denmark.*

102. Rodic, L., & Gupta, S. (2012). Closure and rehabilitation of waste dumpsites in Indian megacities: cases of Delhi and Mumbai. *Proceedings of ISWA World Congress 2012*. Florence, Italy.

103. Rouse, J. (2006). Seeking common ground for people: livelihoods, governance and waste. *Habitat International*, 30(4): 741–753.

104. Sakurai, K. (2000). Municipal Solid Waste Management in Developing Countries: Towards more Efficient International Cooperation in Municipal Solid Waste Management Sector. *Journal of Japan Society of Waste Management Experts*, 11 (2), 142-151.

105. Scheinberg, A., Wilson, D., & Rodic, L. (2010b). *Solid Waste Management in the World's Cities*. London, UK: Earthscan for UN-Habitat.

106. Schu¨ beler, P., Wehrle, K., & Christen, J. (1996). *Conceptual Framework for Municipal Solid Waste Management in Low-income Countries*. St Gallen, Switzerland: SKAT.

107. Shah, H., & Ganguli, S. (2001). *Indian J Community Med*, 63(2), 211-220.

108. Sharma, R., & Mathur, S. (1989). *J Acad Hosp Adm*, (2), 55-57.

109. Sharma, S., & Chauhan, S. (2008). *J. Environ Biol*, 29 (2), 159-162.

110. Stuart, T. (2009). *Waste: Uncovering the Global Food Scandal.* London, UK: Penguin.

111. Tanaka, M. (1996). Health Risk Related to Waste Disposal; For Reductions of Dioxins/Furans. *Journal of National Institute for Public Health.*

112. Thapa, G. (1998). Lessons learned from solid waste management in Kathmandu, Nepal. *Habitat International*, (22): 97-114.

113. Tobin, P., & Myers, C. (1974). Integrated waste management systems. *American Society of Mechanical Engineers.*

114. Van de Klundert, A., & Anschu¨ tz, J. (2001). *Integrated Sustainable Waste Management – The Concept.* Gouda, the Netherlands: WASTE.

115. Veolia Environmental Services. (2007). *Galileo 6, Hazardous Waste Managing Risks, Recovering Materials.*

116. Wagner, T. (2013). Examining the concept of convenient collection: an application to extended producer responsibility and product stewardship frameworks. *Waste Manage*, 33, 499-507.

117. WHO. (1999). *Safe Management of Wastes from Health Care Activities.* WHO.

118. Wilson, D. (2007). Development drivers for waste management. *Waste Management & Research*, 25(3): 198–207.

119. Wilson, D., Rodic, L., Scheinberg, A., & et, a. (2012). Comparative analysis of solid waste management in 20 cities. *Waste Management & Research*, 30(3): 237–254.

120. World Bank . (2012). *Urban Development – Urban Solid Waste Management – Key Obstacles.*

121. Yoshida, F. (1998). *Political Economy of Wastes and Pollutions.* Tokyo, Japan: Iwanami Shoten Publishers.

INDEX